SpringerBriefs in Computer Science

Series Editors

Stan Zdonik
Peng Ning
Shashi Shekhar
Jonathan Katz
Xindong Wu
Lakhmi C. Jain
David Padua
Xuemin Shen
Borko Furht
V. S. Subrahmanian
Martial Hebert
Katsushi Ikeuchi
Bruno Siciliano

For further volumes:
http://www.springer.com/series/10028

Ovidiu Bagdasar

Concise Computer Mathematics

Tutorials on Theory and Problems

 Springer

Ovidiu Bagdasar
University of Derby
Derby
UK

ISSN 2191-5768   ISSN 2191-5776 (electronic)
ISBN 978-3-319-01750-1   ISBN 978-3-319-01751-8 (eBook)
DOI 10.1007/978-3-319-01751-8
Springer Cham Heidelberg New York Dordrecht London

Library of Congress Control Number: 2013947356

Printed on acid-free paper

Springer is part of Springer Science+Business Media (www.springer.com)

Foreword

It is a great pleasure for me to write a short Foreword to this publication which adds to the *SpringerBriefs in Computer Science series*. The work has arisen from the efforts of colleague Dr. Ovidiu Bagdasar in bringing together various topics from past modules and developing material in order to deliver a new module made available for our Year 1 computing students here at the University of Derby. Entitled Computational Mathematics, it introduces fresher students to some fundamental concepts designed to give them a firm mathematical footing on which to base their studies and from which Dr. Bagdasar has produced a programme of tutorials in booklet form here. The chosen areas—Discrete Mathematics, Logic, Linear Algebra and Special Topics—will lead any student through both theory and a series of problems and exercises, consolidating some of the work typically covered in class but also offering independent study options.

Concise Computer Mathematics: *Tutorials on Theory and Problems* should prove to be a valuable resource for undergraduate students of computing in general, and computer science in particular. As an educational tool its aim is to foster knowledge, understanding and improved confidence in a systematic way, and I wish the publication every success.

Derby, July 2013   Peter J. Larcombe

Preface

Overview and Goals

Through the ages, mathematics has established itself as "the door and key to the sciences" (in the words of Roger Bacon, 1214-1294) and since antiquity it represented one of the key areas of interest for human knowledge. However, Euclid is said (following Proclus) to have replied to King Ptolemy's request for an easier way of learning mathematics that "there is no Royal Road to geometry". To most students' despair, Mathematics still seems to be the tumbling stone for anyone trying to understand the world around us. Computer Science students make no exception as they need to master some key areas of mathematics.

This book has its origins in the delivery of a Computational Mathematics module to a body of undergraduate Computing students. The aim of the module was to deliver an easily accessible, self-contained introduction to the basic notions of Mathematics necessary for their degree. The variety of Computing programmes delivered (Computer Science, Computer Games Programming, Computer Forensic Investigation, Computer Networks and Security, Information Technology) posed a real challenge when developing the syllabus for Computational Mathematics. This book reflects the need to quickly introduce students, from a variety of educational backgrounds, to a number of essential mathematical concepts.

Organisation and Features

The material is divided into four units: Discrete Mathematics (Sets, Relations, Functions), Logic (Boolean Types, Truth tables, proofs), Linear Algebra (Vectors, Matrices and Graphics) and Special topics (Graph Theory, Number Theory, Basic elements of Calculus). Each teaching unit was used in conjunction with specifically designed computer-based tests available through the whole semester.

The module was a success as the students engaged well with their work and improved both the attitude towards their studies and their problem solving skills. The flexibility offered by the computer-based tests was highly praised in the feedback gathered from the students.

How to Use the Book

Guidelines for Students

The chapters contain a brief theoretical presentation of the topic, followed by a short further reading list. The problem section is divided into Essential problems (which are direct applications of the theory) and Supplementary problems (which may require a bit more work). Each chapter ends with a Problem Answers section which either provides correct answers or worked solutions for all the problems. Most problems are original, but there are some taken from past teaching materials and whose source is uncertain.

Guidelines for Tutors

Teaching

Each chapter is covered in two hours of lectures and two hours of tutorial.

Assessment

The assessment is based on computer tests and consists of two parts, each worth 50 % of the final mark.

Part I: Four summative online tests covering

- Test 1: Sets, Relations and Functions
- Test 2: Logic and Boolean Types
- Test 3: Vectors, Matrices and Graphics
- Test 4: Graphs, Number Theory, Elements of Calculus.

Each test can be accessed at any time, for multiple times and only the highest grades over all the valid attempts count towards the final mark.

Part II: Two hours summative test with questions from all subjects.

Acknowledgments

First of all, I would like to thank my family for bearing with a part-time husband and father during my first year as a Lecturer. My appreciation goes to my mentors: Richard and Nick—for their excellent guidance and tremendous support through my first steps as an Academic, Peter and Stuart—for being by my side in every step of the way. Finally to all my colleagues at the University of Derby.

June 2013   Ovidiu Bagdasar

Contents

Chapter 1
Sets and Numbers

Abstract Sets are defined as collections of objects of the same kind and are among the fundamental notions in many Mathematical subjects, including Algebra and Calculus. Operations involving sets play a key role in many applications and sets of numbers have been created for solving numerous problems inspired from real life. In this chapter we present some of the key concepts in the theory of sets such as basic set operations, Venn diagrams, set cardinals and motivations for the number systems.

Keywords Sets · Set operations · Venn diagrams · Number systems

1.1 Brief Theoretical Background

Set Definitions and Notations

- set - collection of similar objects (numbers, symbols, etc)
- $x \in A$ - element x is a member of set A
- $x \notin A$ - element x is not a member of set A
- U - universe - contains all the elements of a kind
- $\varnothing$ - empty set - contains no element
- $B \subseteq A$ - inclusion - set B is included or equal to set A (for all $x \in B$, $x \in A$)
- $B \subset A$ (or $B \subsetneq A$) - strict inclusion - B is included, but not equal to A
- $A = B$ - set equality - (sets A and B are equal when $B \subseteq A$ and $A \subseteq B$)
- $\mathcal{P}(A)$ - power set - set of all subsets of A
- $A \times B$ - cartesian product - all pairs (a, b) s.t. $a \in A$ and $b \in B$

O. Bagdasar, *Concise Computer Mathematics*, SpringerBriefs in Computer Science,
DOI: 10.1007/978-3-319-01751-8_1, © The Author(s) 2013

Example: Let $A = \{a, b\}$ and $B = \{a, b, c\}$ be two sets. Then

$$c \in B, c \notin A, A \subset B$$
$$\mathcal{P}(A) = \{\varnothing, \{a\}, \{b\}, \{a, b\}\},$$
$$A \times B = \{(a, a), (a, b), (a, c), (b, a), (b, b), (b, c)\}.$$

Set Operations

For the sets A and B taken from the universe U we define

- $A \cap B = \{x \in U : x \in A \text{ and } x \in B\}$ (intersection)
- $A \cup B = \{x \in U : x \in A \text{ or } x \in B\}$ (union)
- $A \setminus B = \{x \in U : x \in A \text{ and } x \notin B\}$ (difference)
- $A \triangle B = (A \setminus B) \bigcup (B \setminus A)$ (symmetric difference)
- $A^c = (A)^c = \sim (A) = \complement(A) = U \setminus A = \{x \in U : x \notin A\}$ (complement)

Example: For subsets $A = \{1, 2\}$, $B = \{2, 3, 5\}$ of universe $U = \{0, 1, 2, 3, 4, 5\}$ the previously defined operations give

$$A \cap B = \{2\}, \ A \cup B = \{1, 2, 3, 5\}, \ A \setminus B = \{1\}, \ A \triangle B = \{1, 3, 5\}, \ B^c = \{0, 1, 4\}.$$

Venn Diagrams

Set and operations can be represented by Venn diagrams like in Fig. 1.1. The universe U corresponding to two sets (Fig. 1.1a) is divided into four distinct regions, which can be related to the previously defined set operations.

For example, in the above diagrams for two sets (Fig. 1.1a), $A \cap B$ represents region 3, $A \cup B$ regions 1, 2 and 3, $A \setminus B$ region 1, $A \triangle B$ regions 1 and 2, while A^c regions 2 and 4. Venn diagrams are also used to represent operations involving three sets (Fig. 1.1b) or more.

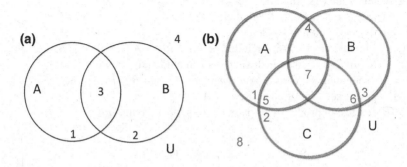

Fig. 1.1 Venn diagrams for **a** two sets and **b** three sets

Ways to Define Sets

$$A = \{0, 1, 2, 3\} \qquad \text{(enumeration)}$$
$$B = \{n \in \mathbb{N} : 3 \leq n \leq 100\} \qquad \text{(ellipsis)}$$
$$C = \{x = n^2 : n \in \mathbb{N}\} \qquad \text{(formula)}$$

Ellipsis should only be used when no confusion is possible.

Other Set Properties

- cardinal - the number of elements in a set.
  Example: The set $\{0, 1, 2\}$ has three elements (therefore cardinal is 3).
- finite vs infinite sets (sets have a finite or infinite number of elements)
  Example: The set $\{0, 1, 2\}$ is finite, while the set $\{0, 1, 2, 3, \ldots\}$ is infinite.
- ordered vs not ordered sets (any two elements are comparable or not)
  Example: The set $\{0, 1, 2\}$ (ordered), the set of students (not ordered).

Number Systems and Motivation

Most people know the natural numbers 0, 1, 2, 3, ..., but other number sets exist. The German mathematician Leopold Kronecker (1850s) was quoted as having said "God made natural numbers; all else is the work of man".

His conclusion was based on the fact that starting with the set of natural numbers, further number sets can be generated by using basic operations (such as subtraction or division of natural numbers), and also by trying to define sets which accommodate the solutions of simple equations.

Basic Number Sets

$$\mathbb{N} = \{0, 1, 2, 3, \ldots\} \qquad \text{natural numbers (including zero)}$$
$$\mathbb{N}^+ = \{1, 2, 3, \ldots\} \qquad \text{natural numbers (excluding zero)}$$
$$\mathbb{Z} = \{\ldots, -2, -1, 0, 1, 2, \ldots\} \qquad \text{integers (positive and negative)}$$
$$\mathbb{Q} = \{q = \tfrac{m}{n} : m \in \mathbb{Z}, n \in \mathbb{N}^+\} \qquad \text{rational numbers (fractions)}$$
$$\mathbb{R} = \{a.b : a, b \in \mathbb{Z}\} \qquad \text{real numbers (infinite decimals)}$$
$$\mathbb{R}^+ = \{x \in \mathbb{R} : x \geq 0\} \qquad \text{positive real numbers}$$
$$\mathbb{C} = \{a + bi : a, b \in \mathbb{R}\} \qquad \text{complex numbers } (i = \sqrt{-1})$$

Polynomials

The most simple equations are generated by polynomials.These are simple mathematical expressions constructed from variables (also called indeterminates) and constants (usually numbers), using only the operations of addition, subtraction, multiplication, and non-negative integer exponents.

Motivation for the Number Systems

Examples of polynomials with roots in the various number sets are

$$
\begin{aligned}
&1.\ x - 1 &&= 0 \Rightarrow x = 1 \in \mathbb{N} \\
&2.\ x + 1 &&= 0 \Rightarrow x = -1 \notin \mathbb{N}, &&x \in \mathbb{Z} \\
&3.\ 4x - 1 &&= 0 \Rightarrow x = 1/4 \notin \mathbb{Z}, &&x \in \mathbb{Q} \\
&4.\ 2x^2 - 1 &&= 0 \Rightarrow x = \sqrt{2} \notin \mathbb{Q}, &&x \in \mathbb{R} \\
&5.\ x^2 + 1 &&= 0 \Rightarrow x = \sqrt{-1} = i \notin \mathbb{R}, &&x \in \mathbb{C}
\end{aligned}
$$

The above examples illustrate the strict inclusions

$$ \mathbb{N} \subsetneq \mathbb{Z} \subsetneq \mathbb{Q} \subsetneq \mathbb{R} \subsetneq \mathbb{C}. $$

The Fundamental Theorem of Algebra (Argand in 1806, Gauss 1816)

Every non-constant single-variable polynomial with complex coefficients has at least one complex root (real coefficients and roots being within the definition of complex numbers).

This theorem shows that the complex numbers are sufficient for solving all polynomial equations (like the ones given in the above examples).

1.2 Essential Problems

E 1.1. What is the power set $\mathcal{P}(A)$ (set of all subsets) for $A = \{a, b, c\}$?

E 1.2. Find the cartesian product of sets $A = \{1, 2, 3, 6\}$ and $B = \{a, b, c\}$ (all pairs whose first component is in A, and second component in B).

E 1.3. (Set operations) Let the sets A, B and C be defined as

$$
\begin{aligned}
A &= \{'r', 'e', 's', 't'\} \\
B &= \{'r', 'e', 'l', 'a', 'x'\} \\
C &= \{'s', 'l', 'u', 'm', 'b', 'e', 'r'\}
\end{aligned}
$$

Find the sets and corresponding regions in the Venn diagram Fig. 1.1b

$$A \cup B \quad A \cup C \quad B \cup C$$
$$A \cap B \quad A \cap C \quad B \cap C$$
$$A \setminus (A \cap C) \quad B \setminus (A \cup C).$$

E 1.4. Consider the Universe $U = \{0, 1, 2, 3, 4, 5, 6, 7\}$ and the sets

$$D = \{0, 1, 2\}, \quad E = \{2, 4, 6\}, \quad F = \{1, 3, 5, 7\}.$$

Find the sets $D \setminus F$, $E \cap F$, $E \cup F$ and $(F \cup D)^c$ and their cardinal.

E 1.5. (a) List the elements of the sets (enumerate or give a formula).

(i) $\{x \in \mathbb{N}^+ : x$ is a multiple of $4\}$

(ii) $\{x \in \mathbb{Z} : -2 \le x \le 3\}$

(b) List the elements of the following sets (enumerate or give a formula).

$$A = \{x \in \mathbb{N}^+ : x \text{ is prime, } x \le 45\}$$
$$B = \{x \in \mathbb{N}^+ : x \text{ is even}\}$$
$$C = \{x \in \mathbb{N}^+ : x \text{ is odd}\}$$
$$D = \{x \in \mathbb{N}^+ : x \text{ is a multiple of } 6\}$$

Are any of the above sets A, B, C or D subsets of one of the other sets?

(c) Determine which of the following sets are equal

$$E = \{0, 1, 2, 3\}$$
$$F = \{x \in \mathbb{N} : x^2 < 13\}$$
$$G = \{x \in \mathbb{N} : (x^2 + 3) < 2\}$$

E 1.6. Solve the following inclusion problems:

(i) If $A \subseteq C$ and $C \subseteq B$, does it follow that $A \subseteq B$?

(ii) If $A \subseteq C$ and $B \subseteq C$, what can you say about A and B?

E 1.7. If A and B are subsets of a universe U, show by constructing examples that each of the following is true.

(i) $(A \triangle B)^c = (A \cap B) \cup (A \cup B)^c$

(ii) $(A \setminus B)^c = A^c \cup B$

E 1.8. Show that the following set identities are true, using Venn diagrams.

    (i) $(A \cap B) \cup (A \cap B)^c = U$(universal set)

    (ii) $(A^c \cap B) \cup A = A \cup B$

1.3 Supplementary Problems

S 1.1. (Cartesian product)

(a) Find $A \times B \times C$ for the sets $A = \{1, 2, 3\}$, $B = \{a, b, c\}$ and $C = \{a, b, c\}$.
(b) If the sets A, B and C have m, n and p $(m, n, p \in \mathbb{N}^+)$ elements respectively, what is the number of elements of $A \times B \times C$?

S 1.2. Prove that if A, B, C are sets from the Universe U, then:

    (i) $A \cap (B \cup C) = (A \cap B) \cup (A \cap C)$ (distributivity of $\cap$)

    (ii) $A \cup (B \cap C) = (A \cup B) \cap (A \cup C)$ (distributivity of $\cup$)

    (iii) $(A \cap B)^c = A^c \cup B^c$ (de Morgan law for $\cap$)

    (iv) $(A \cup B)^c = A^c \cap B^c$ (de Morgan law for $\cup$).

S 1.3. Find the sets A, B with the properties

    (i) $A \cup B = \{1, 2, 3, 4, 5, 6, 7, 8, 9, 10\}$

    (ii) $A \cap B = \{3, 4, 5\}$

    (iii) $A \setminus B = \{2, 6\}$

S 1.4. Write the following numbers as fractions

    (i) $x = 0.333 \cdots = 0.(3) = 0.\overline{3}$

    (ii) $x = 0.23232323 \cdots = 0.(23) = 0.\overline{23}$

    (iii) $x = 0.1129292929 \cdots = 0.11(29) = 0.11\overline{29}$

    (iv) $x = \sqrt{3}$.

1.4 Problem Answers

Essential Problems

E 1.1. $\mathcal{P}(A) = \{\varnothing, \{a\}, \{b\}, \{c\}, \{a, b\}, \{a, c\}, \{b, c\}, \{a, b, c\}\}$.

E 1.2. $A \times B = \{(1, a), (1, b), (1, c), (2, a), (2, b), (2, c), (3, a), (3, b), (3, c), (6, a), (6, b), (6, c)\}$.

E 1.3. The set operations give the following sets

$A \cup B = \{`r`, `e`, `s`, `t`, `l`, `a`, `x`\}$; $A \cup C = \{`r`, `e`, `s`, `t`, `l`, `u`, `m`, `b`\}$;
$B \cup C = \{`r`, `e`, `l`, `a`, `x`, `s`, `u`, `m`, `b`\}$; $A \cap B = \{`r`, `e`\}$;
$A \cap C = \{`r`, `e`, `s`\}$;
$B \cap C = \{`r`, `e`, `l`\}$; $A \setminus (A \cap C) = \{`s`, `t`\}$; $B \setminus (A \cup C) = \{`a`, `x`\}$.

The regions are $A \cup B$ (1, 5, 4, 7, 6, 3), $A \cup C$ (1, 5, 4, 7, 2, 6), $B \cup C$ (2, 5, 6, 7, 4, 3), $A \cap B$ (4, 7), $A \cap C$ (5, 7), $B \cap C$ (6, 7), $A \setminus (A \cap C)$ (1, 4), $B \setminus (A \cup C)$ (3).

E 1.4. $D \setminus F = \{0, 2\}$; $E \cap F = \varnothing$; $E \cup F = \{2, 4, 6, 1, 3, 5, 7\}$; $(F \cup D)^c = \{4, 6\}$. The cardinals are 2, 0, 7 and 2.

E 1.5.   (a) (i)$\{4, 8, 12, \ldots\} = \{4n : n \in \mathbb{N}^+\}$; (ii)$\{-2, -1, 0, 1, 2, 3\}$.

       (b) $A = \{2, 3, 5, 7, 11, 13, 17, 19, 23, 29, 31, 37, 41, 43\}$

         $B = \{2, 4, 6, 8, \ldots\} = \{2n : n \in \mathbb{N}^+\}$

         $C = \{1, 3, 5, 7, \ldots\} = \{2n - 1 : n \in \mathbb{N}^+\} = \{2n + 1 : n \in \mathbb{N}\}$

         $D = \{6, 12, 18, \ldots \mathbb{N}^+\} = \{6n : n \in \mathbb{N}^+\} \subset B$

       (c) $E = F = \{0, 1, 2, 3\}$.

E 1.6. (i) yes;   (ii) nothing; The inclusion can be true or false.
We may have the following situations:

- If $A = \{2n : n \in \mathbb{N}^+\}$, $B = \{2n + 1 : n \in \mathbb{N}^+\}$ and $C = \mathbb{N}$ then $A \subset C$, $B \subset C$ and $A \cap B = \varnothing$ (A and B have no common elements).
- If $A = \{2n : n \in \mathbb{N}^+\}$, $B = \{4n : n \in \mathbb{N}^+\}$ and $C = \mathbb{N}$ then $A \subset C$, $B \subset C$ and $A \cap B = \varnothing$ but $B \subset A$ (B is included in A).

E 1.7. Let $U = \{0, 1, 2, 3, 4, 5, 6, 7, 8, 9\}$ and $A = \{0, 1, 2, 3, 4, 5\}$, $B = \{4, 5, 6, 7\}$.

      (i) $(A \triangle B) = \{0, 1, 2, 3, 6, 7\}$, $(A \cap B) = \{4, 5\}$, $(A \cup B)^c = \{8, 9\}$
          $(A \triangle B)^c = (A \cap B) \cup (A \cup B)^c = \{4, 5, 8, 9\}$
      (ii) $(A \setminus B)^c = A^c \cup B = \{4, 5, 6, 7, 8, 9\}$

E 1.8. The expressions correspond to the same regions of the Venn diagrams.

Supplementary Problems

S 1.1. (b) mnp.

(a) $A \times B \times C = \{(1, a, a), (1, a, b), (1, a, c), (1, b, a), (1, b, b), (1, b, c), (1, c, a),$
$(1, c, b), (1, c, c), (2, a, a), (2, a, b), (2, a, c), (2, b, a), (2, b, b),$
$(2, b, c), (2, c, a), (2, c, b), (2, c, c), (3, a, a), (3, a, b), (3, a, c),$
$(3, b, a), (3, b, b), (3, b, c), (3, c, a), (3, c, b), (3, c, c)\};$

S 1.2. You may use the Venn diagrams to show that both sides determine the same region. Otherwise, we can use the following argument.

(i) Let $x \in A \cap (B \cup C)$ then $x \in A$ and $x \in (B \cup C)$. This means that $x \in A$ and $x \in B$ or $x \in C$. This means that $x \in (A \cap B)$ or $x \in (A \cap C)$. For the proof to be complete one also need to start with a x from the right hand side, and show that it belongs to the set at the left.

(ii) $x \in A \cup (B \cap C) \Leftrightarrow x \in A$ or $x \in (B \cap C)$. If $x \in A$, it clearly belongs to $(A \cup B)$ and $(A \cup C)$, therefore to their intersection. If $x \in (B \cap C)$, it belongs to both B and C, therefore to $(A \cup B)$ and $(A \cup C)$ and $(A \cup B) \cap (A \cup C)$.

(iii) and (iv) - check with Venn diagrams.

S 1.3. One can use a Venn diagram to show that

$$A = \{2, 3, 4, 5, 6\}$$
$$B = \{1, 3, 4, 5, 7, 8, 9, 10\}$$

S 1.4. (i) $x = 0.333 \cdots = 0.(3)$.
Multiplying by 10 we obtain $10x = 3 + x$, which shows that $x = \frac{1}{3}$.

(ii) $x = 0.23232323 \cdots = 0.(23)$.
Multiplying by 100 we obtain $100x = 23 + x$, which shows that $x = \frac{23}{99}$.

(iii) $x = 0.1129292929 \cdots = 0.11(29)$.
Multiplying by 100 the fraction becomes $100x = 11 + 0.(29) = 11 + \frac{29}{99}$.
In the end, we obtain $x = \frac{1129}{9900}$.

(iv) $x = \sqrt{3}$. This is actually not a rational number.

To prove this we assume there are two natural numbers p, q s.t they have no common divisors and $\frac{p}{q} = \sqrt{3}$. This is equivalent to $p^2 = 3q^2$ and because p and q are natural numbers, p^2 is a multiple of 3. This can only happen when p itself is a multiple of 3, therefore $p = 3p_1$. In this case, $(3p_1)^2 = 3q^2$, which gives $3p_1^2 = q^2$. Using the same argument as before, $q = 3q_1$. However, this would mean that 3 is a common divisor of p and q, which is a contradiction. This ends the proof.

Chapter 2
Relations and Databases

Abstract In mathematics a binary relation between two sets A and B is a collection of ordered pairs (a, b) belonging to the cartesian product $A \times B$. In this chapter we present some of the general properties of relations and their operations, as well some special types of relations defined over the same set. A good command of these basic notions is essential for understanding the relational database theory, where a relation is a set of tuples $(a_1, a_2, ..., a_n)$ and each element a_j is a member of A_n a data domain (sometimes called an attribute). In the context of databases we also present the notion of key.

Keywords Binary relations · Operations on relations · Databases · Keys

2.1 Brief Theoretical Background

Notations and Definitions

- A binary relation R is a triplet (X, Y, G), where G is a subset of $X \times Y$
- X - domain, Y - co-domain
- G - graph (we usually identify the relation R with its graph G!!!)
- If the pair (x, y) is in G, x is in relation with y ($(x, y) \in G$, xRy or $x \sim y$).

Representation

A relation is defined by the set of elements or by a diagram.
For the sets $X = \{1, 2, 3, 4\}$ and $Y = \{a, b, c, d\}$ one can define a relation as

$$R = \{(1, a), (1, b), (2, e), (3, b), (3, c), (4, c)\}, \tag{2.1}$$

represented in the diagram (or sometimes in a table).

O. Bagdasar, *Concise Computer Mathematics*, SpringerBriefs in Computer Science, DOI: 10.1007/978-3-319-01751-8_2, © The Author(s) 2013

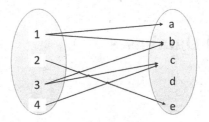

In this relation we can see that $(1, a) \in R$ (or $1Ra$), while $(1, c) \notin R$. At the same time, no element in X is related to $d \in Y$.

Classification of Relations

Based on the number of links between elements of X and Y, a relation is

- many-one: each x related to at most one y
- one-many: each y related to at most one x
- one-one: both many-one and one-many (called functions in Chap. 3)
- neither of the above: this is the case for relation R defined in (2.1).

Operations on Relations

Let X, Y be two sets, $A \subset X$, $B \subset Y$ be subsets, and let $R, S \subset X \times Y$ be two relations. The following operations are defined:

- $R^{-1} = \{(y, x) \in Y \times X : (x, y) \in R\} \subset Y \times X$ (inverse)
- $A \lhd R = \{(x, y) \in X \times Y : (x, y) \in R \text{ and } x \in A\}$ (restriction)
- $R \rhd B = \{(x, y) \in X \times Y : (x, y) \in R \text{ and } y \in B\}$ (co-restriction)
- $R(A) = \{(y \in Y : \text{ there is a pair } (a, y) \in R \text{ with } a \in A\}$ (image - set!!!)
- $R \cup S = \{(x, y) : (x, y) \in R \text{ or } (x, y) \in S\} \subset X \times Y$ (union)
- $R \cap S = \{(x, y) : (x, y) \in R \text{ and } (x, y) \in S\} \subset X \times Y$ (intersection).

Example: For $X = \{1, 2, 3, 4\}$, $Y = \{a, b, c, d\}$, $A = \{1, 2\} \subset X$, $B = \{a, b\} \subset Y$

$$R = \{(1, a), (1, b), (2, e), (3, b), (3, c), (4, c)\},$$
$$S = \{(1, b), (2, a), (3, d), (4, c)\},$$

the above operations give

- $R^{-1} = \{(a, 1), (b, 1), (e, 2), (b, 3), (c, 3), (c, 4)\} \subset Y \times X$
- $A \lhd R = \{(1, a), (1, b), (2, e)\} \subset X \times Y$
- $R \rhd B = \{(1, a), (1, b), (3, b)\} \subset X \times Y$
- $R(A) = \{a, b, e\} \subset Y$ (set!)
- $R \cup S = \{(1, a), (1, b), (2, a), (2, e), (3, b), (3, d), (3, c), (4, c)\} \subset X \times Y$
- $R \cap S = \{(1, b), (4, c)\} \subset X \times Y$.

Composition of Binary Relations

Let $R \subset X \times Y$ and $S \subset Y \times Z$ be two relations.
The composition is a relation $S \circ R \subset X \times Z$ defined as

$$S \circ R = \{(x, z) \in X \times Z : \text{ there exists } y \in Y \text{ with } (x, y) \in R \text{ and } (y, z) \in S\}.$$

These are the pairs (x, z) from $X \times Z$, connected through Y.

Example:
Consider the set $X = \{1, 2, 3, 4\}$ and the relations $R = \{(1, 2), (1, 1), (1, 3), (2, 4), (3, 2)\}$ and $S = \{(1, 4), (1, 3), (2, 3), (3, 1), (4, 1)\}$. Find $S \circ R$.

Solution: As $(1, 2) \in R$ and $(2, 3) \in S$, one has $(1, 3) \in S \circ R$.
Similarly we can proceed with the others:

- $(1, 1) \in R$ and $(1, 4) \in S \Rightarrow (1, 4) \in S \circ R$
- $(1, 1) \in R$ and $(1, 3) \in S \Rightarrow (1, 3) \in S \circ R$
- $(1, 3) \in R$ and $(3, 1) \in S \Rightarrow (1, 1) \in S \circ R$
- $(2, 4) \in R$ and $(4, 1) \in S \Rightarrow (2, 1) \in S \circ R$
- $(3, 2) \in R$ and $(2, 3) \in S \Rightarrow (3, 3) \in S \circ R$

This finally gives $S \circ R = \{(1, 3), (1, 4), (1, 1), (2, 1), (3, 3)\}$.

Special Binary Relations Defined Over a Set

- reflexive: for all $x \in X$ it holds that xRx (other notations: $x \sim y$, $(x, x) \in R$)
- symmetric: For all $x, y \in X$ it holds that if xRy then yRx
- antisymmetric: For all distinct $x, y \in X$, if xRy and yRx then $x = y$
- transitive: For all $x, y, z \in X$, if xRy and yRz then xRz
- total: For all $x, y \in X$, it holds that xRy or yRx
- equivalence: a relation that is reflexive, symmetric and transitive.

Examples:

1. X: all people in the world; xRy: x is a blood relative of y.
   The relation is reflexive, symmetric, transitive (equivalence) and total.
2. $X = \mathbb{N}$; xRy if $x \leq y$.
   The relation is reflexive, antisymmetric, transitive and total.
3. $X = \mathbb{N}$; xRy if $x > y$.
   The relation is not reflexive, not antisymmetric, but transitive and total.

N-ary Relations (Databases) and Keys

Let $n \geq 2$ be a number and $A_1, \ldots, A_n$ sets (attributes). A n-ary relation is a set $R \subset A_1 \times \cdots \times A_n$ of n-tuples $(a_1, \ldots, a_n)$ s.t. $a_j \in A_j, j = 1, \ldots, n$.

- Relation scheme: table with its column headings (called attributes). The entries in the table are called "tuples".
- Key: Set K of attributes for relation R s.t.

  (1) no relation in R can have two tuples with the same values in K, but different on some other attribute,
  (2) no subset of K has a determining property.

Examples: In the following table, the keys are {ID}, {Name}, {Surname, Programme}.

ID	Name	Surname	Programme
11	George	Bush	HND
12	Alan	Turing	PhD
13	Jimmy	Carter	BSc
14	Roger	Smith	MSc
15	Paul	Smith	HND
16	Greg	Smith	Phd

Each entry has four attributes, therefore this is a 4-ary database. The ID and Name attributes can uniquely identify the entries. The surname or the programme cannot uniquely identify the entries on their own, but they form a key when combined.

Remark: In the context of keys one can better understand why students or employees of big companies are given an ID number!

2.2 Essential Problems

E 2.1. Let C be all characters and $Y = \{\text{"apple"}, \text{"pie"}\}$. Describe the relation

$$APPEARS \subset C \times Y$$

where $(c, s) \in APPEARS$ if the character c appears in the string s.

E 2.2. A relation $R \subset A \times B$, where $A = \{1, 2, 3, 4\}$ and $B = \{$ 'a', 'b' , 'c', 'd' $\}$ is shown below. (i) Give R as a table; (ii) Give R^{-1} as a set of pairs.

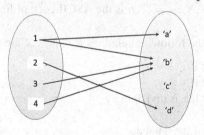

E 2.3. Let $X = \{1, 2, 3, 4\}$, $Y = \{A, B, C, D, E, F\}$ and the relations

$$R \subset X \times Y = \{(1, A), (1, C), (2, A), (3, D), (4, D)\}$$
$$S \subset Y \times Y = \{(E, A), (E, B), (B, D), (C, F), (A, F)\}.$$

Calculate each of the following sets or relations:

(i) $\{1, 2\} \lhd R$ (ii) $S \rhd \{A, C, D\}$ (iii) $R(\{1, 2\})$ (iv) $S \circ R$

(v) $S \circ S$ (vi) $S \circ S \circ S$ (vii) $R \circ R^{-1}$

E 2.4. Suppose that T is a set of students, S is a set of sports, and $LIKES \subset T \times S$ is a relation, as defined below

$$T = \{Anne, Bill, Carol, Dave, Emma, Frankie\}$$
$$S = \{hockey, netball, tennis\}$$
$$LIKES = \{(Anne, tennis), (Anne, hockey), (Bill, tennis), (Carol, netball),$$
$$(Dave, hockey)\}.$$

If $A = \{Bill, Dave, Frankie\} \subset T$ and $B = \{hockey, tennis\} \subset S$ then find $A \lhd LIKES$, $LIKES \rhd B$ and $LIKES\ (A)$.

E 2.5. Let $A \subset X$, $B \subset Y$ be subsets and $R, S \subset X \times Y$ relations. Prove that

$$(A \lhd R) \rhd B = A \lhd (R \rhd B).$$

E 2.6. Let $R \subset X \times Y$, $S \subset Y \times Z$ and $T \subset Z \times U$ be relations. Show that

(i) $(S \circ R)^{-1} = R^{-1} \circ S^{-1}$

(ii) $(T \circ S) \circ R = T \circ (S \circ R).$

E 2.7. For each of the relations given below, decide whether or not it is:
(a) many-one; (b) one-many; (c) one-one; (d) onto.

(i) $R_1 = \{(s,t) \in S \times S : s \neq \text{""}, t$ is string s with its first character deleted$\}$

(ii) $R_2 = \{(c,s) \in C \times S : LEN(s) \geq 3$ and c is the third character of string $s\}$

(iii) $R_3 = \{(s,n) \in S \times N : s \neq \text{""}, n$ is the ASCII code of first character in $s\}$.

E 2.8. Check whether the following relations $R \subset X \times X$ are equivalences

(i) $X = \mathbb{N}$, $(x,y) \in R$ if $x = y$.

(ii) $X = \mathbb{R}$, $(x,y) \in R$ if $x \geq y$.

(iii) $X = \mathbb{Z}$, $(x,y) \in R$ if x is divisible by y.

E 2.9. Find the key(s) of the database

ID	Name	Surname	Programme	Stage
10	Jimmy	Hendrix	BSc	1
12	John	Hendrix	HND	2
16	Jerry	Carter	HND	2

2.3 Supplementary Problems

S 2.1. Let X and Y be sets of cardinalities m and n respectively. What is the minimum cardinality of relation $R \subset X \times Y$? What is the maximal one ?

S 2.2. For $A_1, A_2 \subset X, B_1, B_2 \subset Y$ sets and $R, S \subset X \times Y$ relations prove that

$$(A_1 \cap A_2) \lhd R = (A_1 \lhd R) \cap (A_2 \lhd R)$$
$$(A_2 \cup A_2) \lhd R = (A_1 \lhd R) \cup (A_2 \lhd R)$$
$$R \rhd (B_1 \cap B_2) = (R \rhd B_1) \cap (R \rhd B_2)$$
$$R \rhd (B_1 \cup B_2) = (R \rhd B_1) \cup (R \rhd B_2)$$
$$R(A_1 \cap A_2) = R(A_1) \cap R(A_2)$$
$$R(A_1 \cup A_2) = R(A_1) \cup R(A_2)$$

S 2.3. Let *STUDENTS* be the set of students in a college, *COURSES* be the set of courses the college runs, and *SESSIONS* be the set of sessions (such as: June 21, a.m.) at which exam papers are taken. (Students may take more than one course, and a course may have more than one exam paper.) Relations *TAKING* and *EXAM* are defined as:

TAKING $= \{(s,c) \subset STUDENTS \times COURSES :$ student s is taking course $c\}$

EXAM $= \{(c,d) \subset COURSES \times SESSIONS :$

there is a paper for course c in session $d\}$.

Let F be the set of female students, and C the set of computing courses.

(i) Express in English each of the following:

    (a) $C \lhd EXAM$
    (b) $EXAM(C)$
    (c) $EXAM \circ TAKING$
    (d) $EXAM^{-1} \circ EXAM$
    (e) $EXAM \circ (TAKING \rhd C)$

(ii) Using $F, C, TAKING$ and $EXAM$ and operations on relations, express

    (a) Relation (s, c) where s is a female student taking computing course c.
    (b) The set of courses with no female students.
    (c) The set of exam sessions when the student 'A107X' has a paper to sit.

S 2.4. *HOBBIES* is a relation which is defined below:

$$HOBBIES = \{(Sarah, painting), (Tom, hockey), (Joe, football),$$
$$(Janet, football), (Susie, hockey), (Sarah, sketching),$$
$$(Craig, sketching)\}$$

and also, the following subsets are considered

    $Professional = \{Joe, Sarah\};$ $Amateur = \{Susie, Tom, Janet, Craig\}$
        $Arty = \{sketching, painting\};$ $Games = \{football, hockey\}$

Write down the following:

            (i) $HOBBIES \rhd Arty$
           (ii) $Professional \lhd HOBBIES$
          (iii) $Amateur \lhd HOBBIES \rhd Games$

S 2.5. Let $R \subset X \times Y$ be a binary relation. Express in English the statements:

    (i) $\forall x \in X[\exists y \in Y[(x, y) \in R]];$ (ii) $\forall y \in Y[\exists x \in X[(x, y) \in R]]$

S 2.6. Prove whether the following binary relations are equivalences. If the relation is not an equivalence relation, state why it fails to be one.

        (i) $X = \mathbb{Z} : a \sim b$ if $ab > 0$.
       (ii) $X = \mathbb{R} : a \sim b$ if $a - b$ is rational.
      (iii) $X = \mathbb{R} : a \sim b$ if $a \geq b$.
      (iv) $X = \mathbb{C} : a \sim b$ if $|a| = |b|$.

S 2.7. Find the key(s) of the database

ID	Name	Surname	Programme	Stage
10	Jimmy	Hendrix	BSc	1
11	John	Hendrix	HND	2

2.4 Problem Answers

Essential Problems

E 2.1. *APPEARS =*

C	Y
'a'	"apple"
'p'	"apple"
'l'	"apple"
'e'	"apple"
'p'	"pie"
'i'	"pie"
'e'	"pie"

E 2.2. (i)

1	'a'
1	'b'
2	'd'
3	'b'
4	'b'

(ii) $R^{-1} = \{(\text{'a'},1),(\text{'b'},1), (\text{'b'},3), (\text{'b'},4), (\text{'d'},1)\}$.

E 2.3. The operations give the following results

    (i) $\{1, 2\} \lhd R = \{(1, A), (1, C), (2, A)\}$,

    (ii) $S \rhd \{A, C, D\} = \{(E, A), (B, D)\}$,

    (iii) $R(\{1, 2\}) = \{A, C\}$,

    (iv) $S \circ R = \{(1, F), (2, F)\}$,

(v) $S \circ S = \{(E, F), (E, D)\}$,

(vi) $S \circ S \circ S = \emptyset$,

(vii) $R \circ R^{-1} = \{(A, A), (A, C), (C, A), (C, C), (D, D)\}$.

E 2.4. The answers are

$A \lhd LIKES = \{(Bill, tennis), (Dave, hockey)\}$

$LIKES \rhd B = \{(Anne, tennis), (Anne, hockey), (Bill, tennis), (Dave, hockey)\}$

$LIKES(A) = \{tennis, hockey\}$.

E 2.5. Both sides represent the set

$$(A \lhd R) \rhd B = A \lhd (R \rhd B) = \{(a, b) \in R : a \in A, b \in B\}.$$

E 2.6. Let $R \subset X \times Y$, $S \subset Y \times Z$ and $T \subset Z \times U$ be relations. Show that

(i) $(z, x) \in (S \circ R)^{-1} \Leftrightarrow (x, z) \in (S \circ R)$

$\qquad\qquad\qquad\quad \Leftrightarrow \exists y \in Y$ s.t $(x, y) \in R, (y, z) \in S$

$\qquad\qquad\qquad\quad \Leftrightarrow \exists y \in Y$ s.t $(y, x) \in R^{-1}, (z, y) \in S^{-1}$

$\qquad\qquad\qquad\quad \Leftrightarrow (z, x) \in R^{-1} \circ S^{-1}$

(ii) $(x, t) \in (T \circ S) \circ R \Leftrightarrow \exists y \in Y$ s.t $(x, y) \in R, (y, t) \in T \circ S$

$\qquad\qquad\qquad\quad \Leftrightarrow \exists y \in Y, z \in Z$ s.t $(x, y) \in R, (y, z) \in S, (z, t) \in T$,

$\qquad\qquad\qquad\quad \Leftrightarrow \exists z \in Z$ s.t $(x, z) \in S \circ R, (z, t) \in T$,

$\qquad\qquad\qquad\quad \Leftrightarrow (x, t) \in T \circ (S \circ R)$.

E 2.7. (i) many-one; (ii) one-many; (iii) many-one.

E 2.8.

(i) $X = \mathbb{N}, (x, y) \in R$ if $x = y$. Yes, because it is (a) reflexive: $x = x$; (b) symmetric: $x = y$ implies $y = x$ (c) transitive: $x = y, y = z$ implies $x = z$.

(ii) $X = \mathbb{R}, (x, y) \in R$ if $x \geq y$. Not an equivalence, because it is (a) reflexive: $x \geq x$; (b) **not** symmetric (actually antisymmetric): $x \geq y$ does not imply $y \geq x$ (ex. $x = 3, y = 2$) (c) transitive: $x \geq y, y \geq z$ implies $x \geq z$.

(iii) $X = \mathbb{Z}, (x, y) \in R$ if x is divisible by y. As above, it is not symmetric.

E 2.9. The keys are $\{ID\}$, $\{Name\}$, $\{Surname, Programme, Stage\}$.

Supplementary Problems

S 2.1. $\min = 0$; $\max = mn$.

S 2.2. One just needs to use the definitions of $\cap, \cup, \lhd, \rhd$ and of the image of a subset, and the double inclusion to prove the pairs of sets are equal.

S 2.3. (i) Express in English each of the following:

(a) $C \lhd EXAM = (c, d)$: there is a paper in computing c in the exam session d
(b) $EXAM(C) = $ the sessions d having computing exams
(c) $EXAM \circ TAKING = (s, d)$ the students s taking an exam in session d
(d) $EXAM^{-1} \circ EXAM = (c, e)$ the pairs of courses c,e having a common session
(e) $EXAM \circ (TAKING \rhd C) = (s, d)$ the students s taking computing exam in session d.

(ii)

(a) $F \lhd TAKING \rhd C$;
(b) $TAKING(STUDENTS \setminus F)$;
(c) $EXAM(TAKING(\text{`}A107X\text{'}))$.

S 2.4. The answer is

(i) $HOBBIES \rhd Arty = \{(Sarah, painting), (Sarah, sketching), (Craig, sketching)\}$

(ii) $Professional \lhd HOBBIES = \{(Sarah, painting), (Joe, football), (Sarah, sketching)\}$

(iii) $Amateur \lhd HOBBIES \rhd Games = \{(Janet, football), (Tom, hockey), (Susie, hockey)\}$.

S 2.5. (i) for all x in X, there is y in Y s.t. (x, y) is in R;
(ii) for all y in Y, there is x in X s.t. (x, y) is in R.

S 2.6. The following answers are obtained:

(i) $X = \mathbb{Z} : a \sim b$ if $ab > 0$.
Yes: reflexive, symmetric and transitive.

(ii) $X = \mathbb{R} : a \sim b$ if $a - b$ is rational.
Yes: reflexive, symmetric and transitive.

(iii) $X = \mathbb{R} : a \sim b$ if $a \geq b$.
No: reflexive, **antisymmetric** and transitive.

(iv) $X = \mathbb{C} : a \sim b$ if $|a| = |b|$.
Yes: reflexive, symmetric and transitive.

E 2.7. The keys are {ID}, {Name}, {Programme}, {Stage}.

Chapter 3
Functions

Abstract Functions are very often the central objects of investigation in most fields of modern mathematics. There are many ways to describe or to represent a function. Functions can be defined by a formula or algorithm that indicates how to compute the output for the given input. Sometimes they are given by a picture (called the graph of the function), or by a table of values. In this chapter functions are introduced as relations which link every input from a given set to a unique output from another set. We then present some basic properties, examples and operations involving functions, along with applications based on set numbers, data structures and polynomials.

Keywords Function properties and operations · Datatypes · Polynomials

3.1 Brief Theoretical Background

Notations and Definitions

- Function: Special type of relation s.t. **each element has a unique image**.
- Function elements: X-domain, Y-codomain, $y = R(x) \in Y$ - image of x.

Examples of Useful Functions

- $Card$ (or $\sharp$): $\mathcal{P}(X) \to \mathbb{N}$, $Card(A) =$ the number of elements in subset A
- $d : \mathbb{R} \times \mathbb{R} \to \mathbb{R}^+$, $d(x, y) = |x - y|$ distance between numbers x and y
- Operations on $\mathbb{N} \times \mathbb{N} \to \mathbb{N}$: well defined $(+, \times)$ and ill defined $(-, /)$.

  - $+ : \mathbb{N} \times \mathbb{N} \to \mathbb{N}$, $(x, y) \longmapsto x + y$ (well defined: $x + y \in \mathbb{N}$)
  - $\times : \mathbb{N} \times \mathbb{N} \to \mathbb{N}$, $(x, y) \longmapsto x \times y$ (well defined: $x \times y \in \mathbb{N}$)
  - $- : \mathbb{N} \times \mathbb{N} \to \mathbb{N}$, $(x, y) \longmapsto x - y$ (ill defined: $(1, 3) \longmapsto -2 \notin \mathbb{N}$)
  - $/ : \mathbb{N} \times \mathbb{N} \to \mathbb{N}$, $(x, y) \longmapsto x/y$ (ill defined: $(1, 2) \longmapsto 1/2 \notin \mathbb{N}$)

O. Bagdasar, *Concise Computer Mathematics*, SpringerBriefs in Computer Science, DOI: 10.1007/978-3-319-01751-8_3, © The Author(s) 2013

- power of a given exponent: $f(x) = 2^x$ ($f(1) = 2, f(2) = 4, etc$)
- $\exp x, \log x, \sin x$ and $\cos x$.

General Properties of Functions $f : X \rightarrow Y$

- injective: different elements have different images

$$x_1, x_2 \in X : x_1 \neq x_2 \Rightarrow f(x_1) \neq f(x_2)$$
$$(\text{Example:} f : \mathbb{R} \rightarrow \mathbb{R}, \quad f(x) = -2x + 1)$$

- surjective: all elements in Y are image of elements from X.

$$y \in Y : \text{ there is } x \in X \text{ s.t. } f(x) = y$$
$$(\text{Example:} f : \mathbb{R} \rightarrow \mathbb{R}, \quad f(x) = -2x + 1)$$

- bijective (1-on-1): a function that is both injective and surjective
- a function f only has an inverse f^{-1} if it is bijective!
- A function $f : \mathbb{R} \rightarrow \mathbb{R}$ is increasing (decreasing if $f(x_1) \geq f(x_2)$) if

$$x_1, x_2 \in \mathbb{R} : x_1 \leq x_2 \Rightarrow f(x_1) \leq f(x_2)$$
$$(\text{Example: } f : \mathbb{R} \rightarrow \mathbb{R}, \quad f(x) = 2x + 1)$$

- A function $f : [-a, a]$ or $\mathbb{R} \rightarrow \mathbb{R}$ is even if

$$\text{for all } x \in X : f(x) = f(-x) \quad (\text{Examples:} f(x) = x^2, \cos x)$$

- A function $f : [-a, a]$ or $\mathbb{R} \rightarrow \mathbb{R}$ is odd if

$$\text{for all } x \in X : f(x) = -f(-x) \quad (\text{Examples: } f(x) = x, x^3, \sin x, \tan x).$$

Applications

- Theory of cardinals:

  (1) Two sets have same cardinal if there is a 1–1 function between $X \sim Y$
  (2) Finite sets: have same cardinal is the same number of elements
  (3) Infinite sets: $\mathbb{N} \sim \mathbb{Z} \sim \mathbb{Q}$ (countable), $\mathbb{R}$ (not countable).

- Datatype (set, together with the basic defining functions) string:

  (1) Sets: C set of all characters, S set of all (ordered) sequences of characters
  (2) Notation: 'a' - character, "string" - string

(3) Functions defining the datatype string (all but LEN):

- $LEN : S \rightarrow \mathbb{N}$ - number of characters in string s
- $FIRST : S \rightarrow C$ - selects the first character in string s
  The domain of $FIRST$ is $S \setminus \{$" "$\}$ (i.e. s is not an empty string).
- $REST : S \rightarrow S$ - deletes the first character from string s
  The domain of $REST$ is $S \setminus \{$" "$\}$ (i.e. s is not an empty string)
- $ISEMPTY : S \rightarrow \{true, false\}$ - checks whether a string is empty
- $ADDFIRST : C \times S \rightarrow S$ - adds character c at the front of string s
- $STR : C \rightarrow S$ - makes the string "c" out of the character 'c'.

• Functions ASC and CHR (for characters):

- ASCII code: number between 0 and 255 representing a keyboard character on one byte (8 bit).
- $ASC : C \rightarrow \{0, \dots, 255\}$ - converts a character to its ASCII code
- $CHR : \{0, \dots, 255\} \rightarrow C$ - converts an ASCII code to a character
  Example: $ASC('a') = 97, CHR(98) = 'b'$.

• Special notations and operations

- $+_\mathbb{R}, =_\mathbb{N}, <_C, =_s$: defined on $\mathbb{R}$(real), $\mathbb{N}$(natural), C(char), S(string)
  Example: $=_C$ is well defined when **both** inputs are in set C.
- $>_c$: orders two characters by comparing their ASCII codes.
- $+_s$: concatenates two strings ("a" $+_s$ "xe" $=$ "axe")

Composition of Functions

Let $f \subset X \times Y$ and $g \subset Y \times Z$ be two functions.

$$g \circ f(x) = \{z \in Z : \text{there exists } y \in Y \text{ with } f(x) = y \text{ and } g(y) = z\}.$$

Note that in general $g(f(x)) \neq f(g(x))$!!!

Example: Let the functions $f, g : \mathbb{R} \rightarrow \mathbb{R}$ be $f(x) = 2x + 1, g(x) = x^2 - 1$. Then

$$f(g(x)) = 2 \cdot g(x) + 1 = 2(x^2 - 1) + 1 = 2x^2 - 1,$$
$$g(f(x)) = (f(x))^2 - 1 = (2x + 1)^2 - 1 = 4x^2 + 4x.$$

Polynomials and Polynomial Functions

Polynomials: Simple mathematical expressions constructed from variables (called indeterminates) and constants (usually numbers), using the operations of addition, subtraction, multiplication, and natural exponents.

Examples: The following expressions are polynomials:

$$P = 1 \qquad \qquad \text{(degree 0–constant)}$$
$$Q = 2x - 1 \qquad \qquad \text{(degree 1–linear)}$$
$$R = 4x^2 - 2x + 1 \qquad \qquad \text{(degree 2–quadratic)}$$
$$T(x, y) = x^2 - 2xy + 3y - 1 \qquad \qquad \text{(degree 2 in x and 1 in y)}$$

Operations with Polynomials

Let $P = 2x + y + 1$ and $Q = 3x^2 + 2xy + 4x - 3$. We can define the operations

- multiplication by a scalar

$$2P = 4x + 2y + 2$$

- addition:

$$P + Q = 3x^2 + 2xy + 6x + y - 2$$

- multiplication

$$PQ = (2x + y + 1)(3x^2 + 2xy + 4x - 3)$$
$$= 6x^3 + 7x^2y + 11x^2 + 2xy^2 + 6xy - 2x - 3y - 3.$$

Polynomial function: Function defined by evaluating a polynomial.
Composing two polynomial functions generates a polynomial functions.

Root of a polynomial: A number r is called a root of polynomial P if the value of
the polynomial function at r is zero ($P = 0$).

Example: Polynomial $P = x^2 - 3x + 2 = (x - 1)(x - 2)$ has roots 1 and 2.

3.2 Essential Problems

Hints: Make sure you understand the theory, before solving the problems!

When solving the problems one has to notice that the operations with index
(" $+_s$ ", " $=_c$ ", " $=_\mathbb{N}$ ", " $+_\mathbb{R}$ ") refer to operations on the sets: S (strings),
C (characters), $\mathbb{N}$ (natural numbers) and $\mathbb{R}$ real numbers.
The apostrophe notation 'h' denotes character h, "h" denotes string h.

E 3.1. Calculate $ASC(x) +_\mathbb{N} 32$ for $x = $ 'A' and $x = $ 'p'.

E 3.2. Which of the following are true? If false or invalid explain why.

$$\text{(a) } `a' <_c `A'; \quad \text{(b) } `2' <_c `Q'; \quad \text{(c) } 3 =_c `3'.$$

E 3.3. Which of the following are true, false (or invalid).

(a) $STR(`c') =_s \text{``}c\text{''}$; (b) $\text{``}d\text{''} =_c `d'$; (c) $ASC(`c') \in C$;
(d) $STR(ASC(`b')) =_s \text{``}98\text{''}$.

E 3.4. Given $s = \text{``}Hal\text{''}, t = \text{``}PRINCE\text{''}$ compute where possible:

(a) $LEN(s)$;
(b) $LEN(t)$;
(c) $s +_R t$;
(d) $FIRST(s)$;
(e) $REST(t)$;
(f) $t +_s s$;
(g) $t +_s `s'$;
(h) $REST(s) +_s \text{``}ice\text{''}$.

E 3.5. Find $FIRST(\text{``}12345\text{''})$ ; $LEN(\text{``}take\ away\text{''})$.

E 3.6. Compute where possible, or say if the expression is undefined.

(a) $ASC(`D') +_N ASC(`d')$;
(b) $CHR(6.5)$;
(c) $STR(`H')$;
(d) $STR(ASC(`A'))$;
(e) $REST(\text{``}Louise\text{''})$;
(f) $FIRST(\text{``}THELMA\text{''})$;
(g) $LEN(\text{``}time\text{''} +_s \text{``}out\text{''})$;
(h) $REST(REST(\text{``}frog\text{''}))$;
(i) $FIRST(REST(\text{``}toad\text{''}))$;
(j) $FIRST(REST(\text{``}K\text{''}))$;
(k) $STR(FIRST(\text{``}frog\text{''})) +_s REST(REST(\text{``}Xi\text{''}))$.

E 3.7. Each of the following statements defines an element IMPLICITLY. Say which element is uniquely defined or whether there is no such element and explain why. Are there are many possible elements?

(a) $c \in S$ such that $c+_s \text{``}ed\text{''} =_s \text{``}pushed\text{''}$;
(b) $t \in S$ such that $LEN(t) = 6$;
(c) $r \in S$ such that $LEN(r) = 0$;
(d) $v \in C$ such that $ASC(v) = 109$;
(e) $w \in N$ such that $LEN(w) = 1$.

E 3.8. Let $f, g, h : \mathbb{R} \to \mathbb{R}$ be functions defined as

$$f(x) = x^2 - 1 \quad g(x) = 2x^2 + x + 1 \quad h(x) = 3x - 2.$$

Evaluate (a) (i) $f \circ f$; (ii) $g \circ f$; (iii) $f \circ h$; (iv) $h \circ f$; (v) $g \circ h$.
(b) (i) $2f - 3g + 4h$; (ii) $f \cdot g$; (iii) $h \cdot f$; (iv) $g \cdot h$.

3.3 Supplementary Problems

S 3.1. Find the maximum domain and codomain of the functions $f : X \rightarrow Y$:
(a) $f(x) = x^2$; (b) $f(x) = \sqrt{x}$; (c) $X = [2, 4]$ and $f(x) = x + 2$.

S 3.2. Define the function $CAPITAL\_CONVERT : S \xrightarrow{\cdot} S$ that replaces a string by the same string, with the first letter replaced by the upper case symbol for that letter. i.e. $fish \rightarrow Fish, ada \rightarrow Ada$. Hint: Check the online for the ASCII codes of lower and upper letters.

S 3.3. Which of the following functions is injective, surjective or bijective:

(a) $f : \mathbb{R} \rightarrow \mathbb{R}, \quad f(x) = 2x + 3$;
(b) $f : \mathbb{R} \rightarrow \mathbb{R}, \quad f(x) = 5$;
(c) $f : \mathbb{R} \rightarrow \mathbb{R}, \quad f(x) = x^2 - 2x + 3$;
(d) $f : \mathbb{R} \rightarrow \mathbb{R}^+, \quad f(x) = x^2 - 2x + 1$;
(e) $f : \mathbb{R}^+ \rightarrow \mathbb{R}^+, \quad f(x) = x^2 - 2x + 1$;
(f) $f : \mathbb{N} \rightarrow \mathbb{N}, \quad f(n) = 2n$;
(g) $f : \mathbb{N} \rightarrow \mathbb{Z}, \quad f(n) = n/2$ (n even), and $f(n) = -(n + 1)/2$ (n odd);
(h) $f : \mathbb{N} \rightarrow 3\mathbb{N}, \quad f(n) = 3n$. Prove that $\mathbb{N}$, $2\mathbb{N}$ and $3\mathbb{N}$ have same cardinal.

S 3.4. Check which of the functions below is well defined

(i) $f : \mathbb{R} \rightarrow \mathbb{R}^+, f(x) = (x + 1)^2 - 2$

(ii) $f : \mathbb{R} \rightarrow \mathbb{R}^+, f(x) = 2x - 1$

(iii) $f : \mathbb{R} \rightarrow \mathbb{R}, f(x) = x - 1$

(iv) $f : \mathbb{R} \rightarrow \mathbb{R}^+, f : \mathbb{R} \times \mathbb{R} \rightarrow \mathbb{R}^+, f(x, y) = x - 2y$

(v) $f : \mathbb{N} \times \mathbb{N} \rightarrow \mathbb{N}^+, f(x, y) = 2x - y$

S 3.5. Prove the following identities

$$x^2 \pm 2xy + y^2 = (x \pm y)^2;$$
$$x^2 - y^2 = (x + y)(x - y);$$
$$x^3 \pm y^3 = (x \pm y)(x^2 \mp 2xy + y^2).$$

S 3.6. Prove that every function $f : \mathbb{R} \rightarrow \mathbb{R}$ can be written as the sum of two functions $g, h : \mathbb{R} \rightarrow \mathbb{R}$, such that g is even and h is odd.

3.4 Problem Answers

Essential Problems

E 3.1. $ASC(`A`) = 65$, $ASC(`p`) = 112$, therefore we have 97 and 144.

E 3.2. (a) false; (b) true; (c) invalid.

E 3.3. (a) true; (b) invalid; (c) false; (d) invalid.

E 3.4. (a) 3; (b) 6; (c) invalid; (d) 'H'; (e) "RINCE"; (f) "PRINCEHal";
(g) invalid; (h) "alice".

E 3.5. '1', 9.

E 3.6. (a) $68 + 100 = 168$; (b) invalid; (c) "H"; (d) STR(65) invalid;
(e) "ouise"; (f) "T"; (g) 7; (h) "og" ; (i) "o"; (j) FIRST(" ") - invalid - the function is
only defined for non-empty strings; (k) "f"$+_s$" "$=$ "f".

E 3.7. (a) "push"; (b) any 6 character string; (c) " "; (d) $ASC(`m`) = 109$, so
$v = `m`$; (e) $w \in \mathbb{N}$ is not a string, therefore $LEN(w)$ is invalid.

E 3.8. (a)

   (i) $f \circ f = x^4 - 2x^2$;
   (ii) $g \circ f = 2(x^2 - 1)^2 + (x^2 - 1) + 1 = 2x^4 - 3x^2 + 2$;
   (iii) $f \circ h = (3x - 2)^2 - 1$;
   (iv) $h \circ f = 3(x^2 - 1) - 2 = 3x^2 - 5$;
   (v) $g \circ h = 2(3x - 2)^2 + (3x - 2) + 1 = 18x^2 - 9x + 7$.

   (b)

   (i) $2f - 3g + 4h = -4x^2 + 9x - 13$;
   (ii) $f \cdot g = 2x^4 + x^3 - x^2 - x - 1$;
   (iii) $h \cdot f = 3x^3 - 2x^2 - 3x + 2$;
   (iv) $g \cdot h = 6x^3 - x^2 + x - 2$.

Supplementary Problems

S 3.1. (a) $f(x) = x^2$: If $X = \mathbb{R}$, then $R^+ \subset Y$; (b) $f(x) = \sqrt{x}$: $X \subset R^+$;
(c) $X = [2, 4]$ and $f(x) = x + 2$: $[4, 6] \subset Y$.

S 3.2. The solution is: domain $s \in S \setminus \{``"\}$ and $97 \leq_\mathbb{N} ASC(FIRST(s)) \leq_\mathbb{N} 122$,
codomain: $t = ADDFIRST(CHR(ASC(FIRST(s)) -_\mathbb{N} 32), REST(s))$.
Remark: five other functions $ASC, CHR, FIRST, REST, ADDFIRST$ have all
been used to make the function $CAPITAL\_CONVERT$.

S 3.3. (a) bijective; (b) neither; (c) neither; (d) surjective; (e) bijective; (f) bijective;
(g) bijective; (h) bijective. $\mathbb{N}$, $2\mathbb{N}$ and $3\mathbb{N}$ have the same cardinal.

S 3.4. Functions (iii) and (iv) are well defined. The others are not.

S 3.5. You just have to multiply the brackets.

S 3.6. The two functions (whose sum is $f(x)$) can be defined as

$$g(x) = \frac{1}{2}[f(x) + f(-x)] \text{ (even)}$$
$$h(x) = \frac{1}{2}[f(x) - f(-x)] \text{ (odd)}.$$

Chapter 4
Boolean Algebra, Logic and Quantifiers

Abstract Boolean algebra was introduced in 1854 by George Boole and has been very important for the development of computer science. It operates with variables which have the truth values true and false (sometimes denoted 1 and 0 respectively). In this chapter the main operations of Boolean algebra (conjunction (AND, $\wedge$), disjunction (OR, $\vee$) and negation (NOT, $\neg$)) are defined, and statements involving logical variables are studied with the aid of truth tables and Venn diagrams. Useful formulae involving logical variables are then discussed (De Morgan's laws), along with the existential and universal quantifiers and their negation.

Keywords Boolean variables · Truth tables · Logical operations · Quantifiers

4.1 Brief Theoretical Background

Boolean Values: $\mathbb{B} = \{true, false\}$ (sometimes the set $\{0, 1\}$ is used instead).

Boolean Operations

- $\neg : \mathbb{B} \to \mathbb{B}$ NOT (standard notation, some books may use "$\sim$")
- $\vee : \mathbb{B} \times \mathbb{B} \to \mathbb{B}$ OR
- $\wedge : \mathbb{B} \times \mathbb{B} \to \mathbb{B}$ AND
- $\Rightarrow : \mathbb{B} \times \mathbb{B} \to \mathbb{B}$ IMPLIES
- $\Leftrightarrow : \mathbb{B} \times \mathbb{B} \to \mathbb{B}$ EQUIVALENT

These operations are usually represented using truth tables like below

p	q	$\neg p$	$p \vee q$	$p \wedge q$	$p \Rightarrow q$	$p \Leftrightarrow q$
T	T	F	T	T	T	T
T	F	F	T	F	F	F
F	T	T	T	F	T	F
F	F	T	F	F	T	T

Equivalence Between Boolean Operations and Set Operations

Let A, B be two subsets of universe U and p and q be defined as

$$p(x) \text{ is true if and only if } x \in A$$
$$q(x) \text{ is true if and only if } x \in B$$

- $\neg p(x)$ is true if and only if $x \in A^c = U \setminus A$.
- $p \vee q(x)$ is true if and only if $x \in A \cup B$.
- $p \wedge q(x)$ is true if and only if $x \in A \cap B$.

Definitions

- Sentence: statement that takes a value true or false.
- Proposition: sentence with no variables.
- Predicate: sentence which contains variables.
- Special predicates:

  - Tautology: always true
  - Contradiction: always false

Useful Formulae for Logical Operations

- **De Morgan's Laws**

$$\neg(p \wedge q) = \neg p \vee \neg q$$
$$\neg(p \vee q) = \neg p \wedge \neg q$$

- **Distributive Law**

$$p \wedge (q \vee r) = (p \wedge q) \vee (p \wedge r)$$
$$p \vee (q \wedge r) = (p \vee q) \wedge (p \vee r)$$

- **Associative Law**

$$p \wedge (q \wedge r) = (p \wedge q) \wedge (p \wedge r)$$
$$p \vee (q \vee r) = (p \vee q) \vee (p \vee r)$$

Observation: Applying $\neg$ to an expression has the following effect:

- p becomes $\neg$p (and of course, $\neg$p becomes p)
- $\wedge$ becomes $\vee$, $\vee$ becomes $\wedge$

Example: $\neg(p \wedge (\neg q \vee r)) = \neg p \vee (p \wedge \neg r)$.

Quantifiers and Their Negation

- universal quantifier: $\forall$ (for all)
  Example: $\forall x \in \mathbb{N}[x + 3 \geq 2]$
- existential quantifier: $\exists$ (there exists)
  Example: $\exists x \in \mathbb{N}[x + 3 > 4]$
- By negation $\forall$ becomes $\exists$ (the proposition negates): $\neg(\forall x[p(x)]) \equiv \exists x[\neg p(x)]$
  Example: $\neg(\forall x \in \mathbb{N}[x + 3 \geq 2]) \equiv \exists x \in \mathbb{N}[x + 3 < 2]$
- By negation $\exists$ becomes $\forall$ (the proposition negates): $\neg(\exists x[p(x)]) \equiv \forall x[\neg p(x)]$
  Example: $\neg (\exists x \in \mathbb{N}[x - 2 > 4]) \equiv \forall x \in \mathbb{N}[x - 2 \leq 4]$

4.2 Essential Problems

E 4.1. Write the negations for each of the following statements

(a) John is six feet tall and he weighs at least 200 pounds.
(b) The bus was late or Tom's watch was slow.

E 4.2. Construct a truth table for the expressions: (i) $\neg(\neg p \wedge q)$; (ii) $(p \vee \neg q) \wedge r$;

(iii) Construct a truth table for $(p \wedge q) \vee r$ and for $(p \vee r) \wedge (q \vee r)$ hence show that these two statements are logically equivalent.
(iv) Check whether $(p \vee q) \wedge \neg r$ and $(p \wedge \neg r) \vee (q \wedge \neg r)$ are equivalent.

E 4.3. For $x \in \mathbb{R}$, suppose that p, q and r are the propositions given below:

$$p: \quad x = -4$$
$$q: \quad x = +4$$
$$r: \quad x^2 = 16$$

Which of the following are true?

(a) $p \Rightarrow r$;   (b) $r \Rightarrow p$;   (c) $q \Rightarrow r$;   (d) $r \Rightarrow q$;   (e) $p \Leftrightarrow r$;

(f) $q \Leftrightarrow r$;   (g) $(p \vee q) \Rightarrow r$;   (h) $r \Rightarrow (p \vee q)$;   (i) $(p \vee q) \Leftrightarrow r$;

E 4.4. Decide which of the sentences below, are universally true, which are universally false, and which are neither. Then write the predicates below in symbolic form:
(i) $(x + 3)^2 > 8$, $x \in \mathbb{N}$;   (ii) $2x = 9$, $x \in \mathbb{N}$.

E 4.5. Check (using truth tables) which statements are logically equivalent

> (i) $(a \wedge b) \vee b$ and $a \wedge b$
>
> (ii) $(a \wedge b) \vee a$ and a
>
> (iii) $(\neg a \wedge b \wedge c) \vee ((a \wedge b) \vee c)$ and $b \wedge c$
>
> (iv) $a \Rightarrow b$ and $\neg b \Rightarrow \neg a$.

E 4.6. Below, y is a variable of type **Nat**, c is of type **Char** and s and t are of type **Str**. Represent the following conditions formally, using only standard functions of the data types **Nat**, **Char**, **Str** and $\wedge, \vee, \neg, \Rightarrow, \Leftrightarrow$.

(i) y is either less than 3, or greater than 6 but less than 10. y is not seven.
(ii) The first character of s is 'b' and the second character of s is 'a'.
(iii) If s starts with 'c' then t starts with 'c'.
(iv) s and t have similar starting characters, but their second characters are distinct.
(v) c is an upper case vowel, and the first two characters of t are the same if and only if the first two characters of s are the same.
(vi) s is the string "fred" or y has the value 16, but not both.

E 4.7. Use quantifiers to express formally the following predicates:

(i) There is a natural number whose square is greater than 7
(ii) Whatever real number x, $x^2 = -(x^2 + 1)$ is false.

E 4.8. Write in English the negations of the following statements

(i) All dogs bark;
(ii) Some birds fly;
(iii) No cat likes to swim;
(iv) No computer science student does not know mathematics.

E 4.9. Find the negation of the following statements and evaluate their truth values:

(i) $\forall x \in \mathbb{R} [x^2 - 6x + 10 \geq 0]$;   (ii) $\forall x \in \mathbb{R} [x^2 \geq x]$;
(iii) $\exists x \in \mathbb{N}^+ [x^2 + 2x - 2 \leq 0]$;   (iv) $\exists x \in \mathbb{Q} [x^2 = 3]$.

4.3 Supplementary Problems

S 4.1. Decide whether each of the following propositions is true or false:

(i) $\sqrt{27} \geq 5$;   (ii) $3^2 + 2^3 < \sqrt{250}$;   (iii) $3^4 < 4^3$.

S 4.2. Repeat question E 4.4. for $x \in \mathbb{R}$.

S 4.3. Decide whether each of the following predicates is true or false.

(i) $\forall x \in \mathbb{R}[x^2 >_\mathbb{R} 0]$;   (ii) $\exists t \in \mathbb{N}[t^2 =_\mathbb{N} 7]$;   (iii) $\exists s \in \mathbb{N}[s^2 =_\mathbb{N} 9]$.

S 4.4. Decide which of the sentences below are universally true in $\mathbb{R}$, which are universally false and which are neither.

(i) $(x + 3)^2 > 6$;   (ii) $x = x + 1$;   (iii) $2x = 7$;   (iv) $(x - 1)^2 = x^2 - 2x + 1$.

S 4.5. Let R be the proposition "Roses are red" and B be the proposition "Violets are blue". Express the following propositions as logical expressions:

(a) If roses are not red, then violets are not blue;
(b) Roses are red or violets are not blue;
(c) Either roses are red or violets are blue (but not both);

Use a truth table to show that (a) and (b) are equivalent.

S 4.6. Interpret the following quantified propositions in English and state which are true and which are false:

(i) $\forall s \in S [(\neg(s = \text{" "})) \Leftrightarrow (LEN(s) \geq 1)]$;
(ii) $\forall s \in S [\exists \text{ 'c'} \in C [c = FIRST(s)]]$;
(iii) $\forall s \in S [(s = \text{" "}) \vee (\exists \text{ 'c'} \in C \text{ s.t. 'c'} = FIRST(s))]$.

4.4 Problem Answers

Essential Problems

E 4.1. John is either not six feet tall, or he weighs less than 200 pounds.

E 4.2. In each case, one needs to build the table for 2 and 3 variables. The final expressions are called H or G (to avoid wide tables).

(i) Let H$= \neg(\neg p \wedge q)$. The corresponding truth table is

p	¬p	q	¬p∧q	H
1	0	1	0	1
1	0	0	0	1
0	1	1	1	0
0	1	0	0	1

(ii) Let H$=(p \vee \neg q) \wedge r$. The corresponding truth table is

p	q	r	¬q	p∨¬q	H
1	1	1	0	1	1
1	1	0	0	1	0
1	0	1	1	1	1
1	0	0	1	1	0
0	1	1	0	0	0
0	1	0	0	0	0
0	0	1	1	1	1
0	0	0	1	1	0

(iii) Let $G=(p \wedge q) \vee r$ and $H=(p \vee r) \wedge (q \vee r)$. The truth table for G and H is

p	q	r	p∧q	p∨r	q∨r	G	H
1	1	1	1	1	1	1	1
1	1	0	1	1	1	1	1
1	0	1	0	1	1	1	1
1	0	0	0	1	0	0	0
0	1	1	0	1	1	1	1
0	1	0	0	0	1	0	0
0	0	1	0	1	1	1	1
0	0	0	0	0	0	0	0

hence the two statements are logically equivalent.

(iv) Let $G=(p \vee q) \wedge \neg r$ and $H=(p \wedge \neg r) \vee (q \wedge \neg r)$.
The combined truth table for G and H is
hence the two statements are logically equivalent.

p	q	r	¬r	p∨q	p∧¬r	q∧¬r	G	H
1	1	1	0	1	0	0	0	0
1	1	0	1	1	1	1	1	1
1	0	1	0	1	0	0	0	0
1	0	0	1	1	1	0	1	1
0	1	1	0	1	0	0	0	0
0	1	0	1	1	0	1	1	1
0	0	1	0	0	0	0	0	0
0	0	0	1	0	0	0	0	0

E 4.3. (a) T;   (b) F;   (c) T;   (d) F;   (e) F;   (f) F;   (g) T;   (h) T;   (i) T;
E 4.4. (i) UT; (ii) UF.
E 4.5. (i) yes;   (ii) yes;   (iii) no;   (iv) yes.

E 4.6. (i) $((y < 3) \vee ((y < 10) \wedge (y > 6))) \wedge \neg(y = 6)$;
   (ii) $FIRST(s) = \text{`}b\text{'} \wedge FIRST(FIRST(s)) = \text{`}a\text{'}$;
   (iii) $(FIRST(s) = \text{`}c\text{'}) \Rightarrow FIRST(t) = \text{`}c\text{'}$;

(iv) $FIRST(s) = FIRST(t) \wedge \neq (FIRST(FIRST(s)) = FIRST(FIRST(t)))$;

(v) $\left(65 \leq ASC(c) \leq 90 \wedge FIRST(t) = FIRST(FIRST(t)) \right) \Leftrightarrow$
$FIRST(s)$
$= FIRST(FIRST(s))$;

(vi) $(s = \text{"}fred\text{"} \vee y = 16) \wedge \neg(s = \text{"}fred\text{"} \wedge y = 16)$.

E 4.7. (i) $\exists x \in \mathbb{N}$ s.t. $x^2 > 7$;   (ii) $\forall x \in \mathbb{R}, x^2 \neq -(x^2 + 1)$.

E 4.8. (i) There is at least one dog which can't bark;
(ii) There is at least one bird which can't fly;
(iii) There is at least one cat which likes to swim;
(iv) There is at least one computer science student who knows mathematics.

E 4.9. (i) $\exists x \in \mathbb{R} [x^2 - 6x + 10 < 0]$ F;   (ii) $\exists x \in \mathbb{R} [x^2 < x]$ T;
(iii) $\forall x \in \mathbb{N}^+ [x^2 + 2x - 2 > 0]$ T;   (iv) $\forall x \in \mathbb{Q} [x^2 \neq 3]$ T.

Supplementary Problems

S 4.1. (i) T;   (ii) F;   (iii) F.

S 4.2. (i) UT; (ii) neither.

S 4.3. (i) F;   (ii) F;   (iii) T.

S 4.4. (i) UT;   (ii) UF;   (iii) neither;   (iv) UT.

S 4.5. (a) $\neg R \Rightarrow \neg B$;   (b) $R \vee \neg B$;   (c) $(R \vee B) \wedge \neg (R \wedge B)$

S 4.6. (i) For all strings s, the propositions "the string is not void" and "the string's length is at least one" are equivalent. True.
(ii) Every string s has a first character 'c'.
False, as it doesn't apply for the empty string s=" ".
(iii) Every string s is either empty, or has a first character 'c'. True.

Chapter 5
Normal Forms, Proof and Argument

Abstract In boolean logic, logical formulae are usually reduced to standardised (or normalised) forms, which are more appropriate for automated theorem proving. Such an example are the disjunctive normal form (d.n.f), represented by conjunctive clauses connected by the disjunction operator (OR, $\vee$), and the conjunctive normal form (c.n.f), represented by disjunctive clauses connected by the conjunction operator (AND, $\wedge$). In this Chapter we present the formal definition and derivation of d.n.f and c.n.f, along with some deduction rules. We then present some popular arguments such as the proof by contradiction, the proof by induction and the pigeonhole principle.

Keywords Normal forms (d.n.f, c.n.f) · Deduction rules · Special arguments

5.1 Brief Theoretical Background

In this chapter we use the boolean variables $\{0, 1\}$.

Disjunctive Normal Form (d.n.f)

Equivalent way of writing the truth table, first using $\wedge$, then using $\vee$.

Example 1: Find the d.n.f of $\neg(\neg p \wedge q)$ (Fig. 5.1).

Step 1: Truth table

p	$\neg p$	q	$\neg q$	$\neg p \wedge q$	$\neg(\neg p \wedge q)$
1	0	1	0	0	1
1	0	0	1	0	1
0	1	1	0	1	0
0	1	0	1	0	1

O. Bagdasar, *Concise Computer Mathematics*, SpringerBriefs in Computer Science, DOI: 10.1007/978-3-319-01751-8_5, © The Author(s) 2013

Fig. 5.1 d.n.f using Truth
tables and Venn diagrams

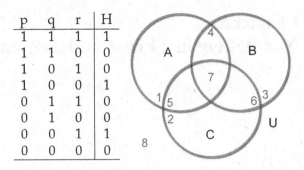

p	q	r	H
1	1	1	1
1	1	0	0
1	0	1	0
1	0	0	1
0	1	1	0
0	1	0	0
0	0	1	1
0	0	0	0

Step 2: Identify 1 entries in the last column.
Step 3: For each entry, the 1s in the first 4 columns are connected by $\wedge$.
Solution: $d.n.f(\neg(\neg p \wedge q)) = (p \wedge q) \vee (p \wedge \neg q) \vee (\neg p \wedge \neg q)$

Example 2: For $H: d.n.f(H) = (p \wedge q \wedge r) \vee (p \wedge \neg q \wedge \neg r) \vee (\neg p \wedge \neg q \wedge r)$
The corresponding regions in the Venn diagram are: 7, 1, 2.

Conjunctive Normal Form (c.n.f)

Conjunctive normal form (c.n.f.) combines the basic literals, first using $\vee$, then
using $\wedge$.
Algorithm for finding the c.n.f of a formula H:

Step 1: Find the d.n.f of $\neg$H
Step 2: Find the negation of this d.n.f

Observation 1: The expressions d.n.f(H) and c.n.f(H) have the same truth value
(maybe less evident for c.n.f, where which is basically equivalent to $\neg(\neg H)$) as they
are just different ways of writing H.

Example 3: Find the c.n.f of $H = \neg p \wedge q$.

Step 1: Truth table and d.n.f of $\neg$H

p	q	$\neg p$	H	$\neg H$
1	1	0	0	1
1	0	0	0	1
0	1	1	1	0
0	0	1	0	1

$$d.n.f(\neg H) = (p \wedge q) \vee (p \wedge \neg q) \vee (\neg p \wedge \neg q).$$

Step 2: As c.n.f(H) is the negation of d.n.f(¬H) one obtains

$$c.n.f(H) = \neg d.n.f(\neg H) = \neg\left[(p \wedge q) \vee (p \wedge \neg q) \vee (\neg p \wedge \neg q)\right]$$
$$= (\neg p \vee \neg q) \wedge (\neg p \vee q) \wedge (p \vee q)$$

Example 4: For the variable H on the previous page, $d.n.f(\neg H)$ is

$$d.n.f(\neg H) = (p \wedge q \wedge \neg r) \vee (p \wedge \neg q \wedge r) \vee (\neg p \wedge q \wedge r) \vee (\neg p \wedge q \wedge \neg r)$$
$$\vee (\neg p \wedge \neg q \wedge \neg r)$$
$$c.n.f(H) = (\neg p \wedge \neg q \wedge r) \vee (\neg p \wedge q \wedge \neg r) \vee (p \wedge \neg q \wedge \neg r)$$
$$\vee (p \wedge \neg q \wedge r) \vee (p \wedge q \wedge r)$$

Deduction Rules

Rules defining a valid argument from a fallacious argument.
When the conjunction of all the assumptions implies the conclusion (this can be checked using truth tables), the argument is valid.

Proof of Modus Ponens

If both p and $p \Rightarrow q$ are true (circled), and also q is true (underlined), then

p	q	$p \Rightarrow q$	$\neg q$	$\neg p$
①	1	①	0	0
1	0	0	1	0
0	1	1	0	1
0	0	1	0	1

the argument is valid!

Common Deduction Rules

The other common deduction rules are summarised in the table below.

Valid	Fallacious
Modus Ponens	Converse error
$p \Rightarrow q; p$	$p \Rightarrow q; q$
q	p
Modus Tollens	Inverse error
$p \Rightarrow q; \neg q$	$p \Rightarrow q; \neg p$
$\neg p$	$\neg q$

Special Types of Argument

Proof by Contradiction
This argument is based on Modus Tollens.

Example. Prove that $\sqrt{2}$ is not rational (fraction).

Proof. Assume that $\sqrt{2}$ **is rational**. Then there are two natural numbers p, q s.t they have no common divisors and $\frac{p}{q} = \sqrt{2}$, therefore $p^2 = 2q^2$. Because p and q are natural numbers, p^2 is a multiple of 2, so p itself is a multiple of 2 ($p = 2p_1$). In this case, $(2p_1)^2 = 2q^2$, which gives $2p_1^2 = q^2$. Using the same argument as before, $q = 2q_1$. We have just proved that 2 is a common divisor of p and q, which is a **contradiction**.

The Pigeonhole Principle
If n items are put into m pigeonholes with $n > m$, then at least one pigeonhole must contain more than one item. This simple theorem has numerous applications in Mathematics and Computer Science.

Example. (Hand-shaking problem) In a room there are $n > 1$ people who can shake hands with one another. Show that there is always a pair of people who will shake hands with the same number of people.

Proof. Each person can shake hands with anybody from 0 to $n - 1$ other people, therefore we have $n - 1$ possible 'holes', because either the '0' or the '$n - 1$' hole must be empty (we can't have persons shaking hands with no one, if there is a person who shook hands with everyone). This leaves n people to be placed in at most $n - 1$ non-empty holes, ensuring duplication.

5.2 Essential Questions

E 5.1. Find the d.n.f. of the following formulae:
(i) $\neg(\neg p \wedge q)$;   (ii) $\neg(\neg p \vee q)$;   (iii) $p \wedge (q \vee r)$.

E 5.2. Find the d.n.f and c.n.f of the following statements
(i) $p \vee (q \wedge r)$;   (ii) $\neg((p \wedge q) \vee r)$;   (iii) $p \Leftrightarrow (q \wedge r)$.

E 5.3. Given the following truth table for a formula H, find the disjunctive normal form for H and hence find the conjunctive normal form of H (Fig. 5.2).

p	$\neg p$	q	$\neg q$	H
1	0	1	0	1
1	0	0	1	0
0	1	1	0	0
0	1	0	1	1

Fig. 5.2 d.n.f using Truth
tables and Venn diagrams

p	q	r	G
1	1	1	0
1	1	0	1
1	0	1	0
1	0	0	0
0	1	1	0
0	1	0	1
0	0	1	1
0	0	0	1

E 5.4. Find d.n.f(G) and c.n.f(G) for formula G in the truth table below and identify the corresponding regions in the Venn diagram.

E 5.5. Use truth tables to prove the arguments Modus Ponens, Modus Tollens, Inverse Error and Converse Error.

E 5.6. Prove that if the product ab is even, then either a or b has to be even.

E 5.7. In a box there are 3 blue balls, 5 red balls and 7 yellow balls. Calculate the minimum number of balls needed to be drawn from the box before
(i) a pair of the same color can be made;
(ii) pairs of each color are made.

5.3 Supplementary Questions

S 5.1. Find the d.n.f. and c.n.f of the following formulae:
(i) $\neg p \wedge (q \Rightarrow p)$;   (ii) $\neg p \wedge (\neg q \vee r)$;   (iii) $\neg p \wedge (q \vee r)$.

S 5.2. Find the d.n.f and c.n.f of the following expressions:
(i) $\neg(p \Rightarrow q)$;   (ii) $p \Rightarrow (q \wedge r)$;   (iii) $\neg(p \Leftrightarrow q)$.

S 5.3. Find the c.n.f of the following formulae:
(i) $\neg p \wedge (q \Rightarrow p)$;   (ii) $\neg p \wedge (\neg q \vee r)$;   (iii) $\neg p \wedge (q \vee r)$.

S 5.4. Use the deduction rules discussed before to prove which of the following arguments are valid and which are fallacious (use truth tables).

(i) If the apple is ripe, then it will be sweet;
The apple is sweet.
Therefore the apple is ripe.

(ii) If I go to the pub, I won't finish my revision;
      If I don't finish my revision, I won't do well in the exam tomorrow.
      Thus if I go to the pub, I won't do well in the exam tomorrow.

(iii) Gill is playing rugby if Tom is not in class;
      If Gill is not playing rugby then Tom will be in class.
      Therefore Tom is not in class or Gill is not playing rugby.

(iv) Sam is studying maths or Sam is studying economics;
      Sam is required to take logic if he is studying maths.
      So Sam is an economics student or he is not required to take logic.

(v)   English men wear bowler hats;
      The man is wearing a bowler hat.
      Therefore he must be English.

S 5.5. Comment on the following argument.
Storing on floppy disk is better than nothing.
Nothing is better than a hard disk drive.
Therefore, storing on floppy disk is better than a hard disk drive.

S 5.6. Prove that if $a, b \in \mathbb{Z}$, then $a^2 - 4b \neq 3$.

S 5.7. Prove that if six distinct numbers are selected from $\{1, 2, \ldots, 9\}$, then some two of these numbers sum to 10.

5.4 Problem Answers

Essential Problems

E 5.1. We first find the truth table for each expression.
Note: The truth tables for (i) and (ii) are combined to save time and space.
Let $G = \neg(\neg p \wedge q)$ and $H = \neg(\neg p \vee q)$. The truth table for G and H is

p	$\neg p$	q	$\neg q$	$\neg p \wedge q$	$\neg p \vee q$	G	H
1	0	1	0	0	1	1	0
1	0	0	1	0	0	1	1
0	1	1	0	1	1	0	0
0	1	0	1	0	1	1	0

One obtains (i) $(p \wedge q) \vee (p \wedge \neg q) \vee (\neg p \wedge \neg q)$;   (ii) $p \wedge \neg q$;
(iii) $(p \wedge q \wedge r) \vee (p \wedge \neg q \wedge r) \vee (p \wedge q \wedge \neg r)$.

E 5.2. We first need to find the truth table for the expressions
$A = p \vee (q \wedge r)$, $B = \neg((p \wedge q) \vee r)$ and $C = p \Leftrightarrow (q \wedge r)$ and their negations.

p	q	r	$q \wedge r$	$p \wedge q$	$(p \wedge q) \vee r$	A	B	C	$\neg A$	$\neg B$	$\neg C$
1	1	1	1	1	1	1	0	1	0	1	0
1	1	0	0	1	1	1	0	0	0	1	1
1	0	1	0	0	1	1	0	0	0	1	1
1	0	0	0	0	0	1	1	0	0	0	1
0	1	1	1	0	1	1	0	0	0	1	1
0	1	0	0	0	0	0	1	1	1	0	0
0	0	1	0	0	1	0	0	1	1	1	0
0	0	0	0	0	0	0	1	1	1	0	0

(i) Using the columns corresponding to A and ¬A we obtain

$$\text{d.n.f(A)} = (p \wedge q \wedge r) \vee (p \wedge q \wedge \neg r) \vee (p \wedge \neg q \wedge r) \vee (p \wedge \neg q \wedge \neg r)$$
$$\vee (\neg p \wedge q \wedge r).$$
$$\text{c.n.f(A)} = \neg \text{d.n.f}(\neg A) = (p \vee \neg q \vee r) \wedge (p \vee q \vee \neg r) \wedge (p \vee q \vee r).$$

(ii) Using the columns corresponding to B and ¬B we obtain

$$\text{d.n.f(B)} = (p \wedge \neg q \wedge \neg r) \vee (\neg p \wedge q \wedge \neg r) \vee (\neg p \wedge \neg q \wedge \neg r) \quad \text{and}$$
$$\text{c.n.f(B)} = (\neg p \vee \neg q \vee \neg r) \wedge (\neg p \vee \neg q \vee r) \wedge (\neg p \vee q \vee \neg r) \wedge (p \vee \neg q \vee \neg r)$$
$$\wedge (p \vee q \vee \neg r).$$

(iii) Using the columns corresponding to C and ¬C we obtain

$$\text{d.n.f(C)} = (p \wedge q \wedge r) \vee (\neg p \wedge q \wedge \neg r) \vee (\neg p \wedge \neg q \wedge r) \vee (\neg p \wedge \neg q \wedge \neg r).$$
$$\text{c.n.f(C)} = (\neg p \vee \neg q \vee \neg r) \wedge (\neg p \vee q \vee \neg r) \wedge (\neg p \vee q \vee r) \wedge (p \vee \neg q \vee \neg r).$$

E 5.3. As seen above, we add the column corresponding to ¬H and obtain
$\text{d.n.f(H)} = (p \wedge q) \vee (\neg p \wedge \neg q)$; $\text{c.n.f(H)} = \neg \text{d.n.f}(\neg H) = (\neg p \vee q) \wedge (p \vee \neg q)$.

E 5.4. As seen above, we add the column corresponding to ¬G and obtain

$$\text{d.n.f(G)} = (p \wedge q \wedge \neg r) \vee (\neg p \wedge q \wedge \neg r) \vee (\neg p \wedge \neg q \wedge r) \vee (\neg p \wedge \neg q \wedge \neg r).$$
$$\text{c.n.f(G)} = (\neg p \vee \neg q \vee r) \wedge (\neg p \vee q \vee \neg r) \wedge (\neg p \vee q \vee r) \wedge (p \vee \neg q \vee \neg r).$$

E 5.5. Check if the conjunction of all assumptions implies the conclusion.
Modus Tollens: If both ¬q and p⇒q are true (circled), then ¬p is true.

p	q	$p \Rightarrow q$	$\neg q$	$\neg p$
1	1	1	0	0
1	0	0	1	0
0	1	1	0	1
0	0	①	①	1

The argument is valid!

Converse error: If both q and p⇒q are true (circled), then p is true or false!

p	q	$p \Rightarrow q$	$\neg q$	$\neg p$
1	①	①	0	0
1	0	0	1	0
0	①	①	0	1
0	0	1	1	1

The argument is fallacious, because of line 3 (true does not imply false)!
The proof for Inverse error is similar and left for the reader.

E 5.6. Suppose that none of a or b is even, therefore there exist $c, d \in \mathbb{Z}$ s.t. $a = 2c+1$ and $b = 2d + 1$, whose product is $ab = (2c + 1)(2d + 1) = 4cd + 2c + 2d + 1$. As the latter number is odd, and there was no error in the proof, the only problem seems to have been the initial assumption. This proves that either a or b is even.

E 5.7. (i) For this to happen one can split the balls into 3 holes containing balls of the same colour. We surely have a repetition if we extract 4 balls.

    (ii) In worst case scenario all the balls of two colours have been extracted, but just one of the remaining colour. The minimum number of balls is then $7 + 5 + 2 = 16$.

Supplementary Problems

S 5.1–S 5.3. The solutions are similar to **E 5.1–E 5.4**.

S 5.4. The arguments are as follows:

(i) Defining p: "apple is ripe" and q: "apple is sweet" the proposition becomes

$$\frac{p \Rightarrow q; \, p}{q} \quad \text{(Modus Ponens, true)}$$

(ii) For p: "go to the pub", q: "finish my revision" and r: "do well in the exam" we have

$$\frac{p \Rightarrow \neg q; \, \neg q \Rightarrow \neg r; \, p}{p \Rightarrow \neg r} \quad \text{(true)}$$

(iii) For p: "Gill is playing rugby", q: "Tom is in class" we have

$$\frac{\neg q \Rightarrow p; \neg p \Rightarrow q}{\neg q \vee \neg p} \quad \text{(false)}$$

(iv) For p: "Sam is studying maths", q: "Sam is studying economics" and r: "Sam is required to take logic" the argument is written as

$$\frac{p \vee q; p \Rightarrow r}{q \vee \neg r} \quad \text{(false)}$$

(v) For p: "man is English", q: "man wars bowler hat" we have

$$\frac{p \Rightarrow q; q}{p} \quad \text{(Converse Error, false)}$$

S 5.5. Storing on floppy disk is better than nothing.
Nothing is better than a hard disk drive.
Therefore, storing on floppy disk is better than a hard disk drive.

S 5.6. Suppose this proposition is false. This conditional statement being false means there exist numbers a and b for which $a, b \in \mathbb{Z}$ is true but $a^2 - 4b \neq 3$ is false. Thus there exist integers $a, b \in \mathbb{Z}$ for which $a^2 - 4b = 3$, or equivalently $a^2 = 4b + 3$ (odd). This shows that a is odd, so there is $c \in \mathbb{Z}$ s.t. $a = 2c + 1$. From the equation we obtain that $a^2 = (2c + 1)^2 = 4c^2 + 4c + 1 = 4b + 3$, which can be written as $c^2 + c - b = \frac{1}{2}$.

However, this is a contradiction as $c^2 + c - b$ is an integer. As all the logic after the first line of the proof is correct, it must be that the first line was incorrect. In other words, we were wrong to assume the proposition was false. Thus the proposition is true.

S 5.7. Since six numbers have to be selected, the Pigeonhole Principle guarantees that at least two of them are selected from one of the five sets $\{1, 9\}$, $\{2, 8\}$, $\{3, 7\}$, $\{4, 6\}$, $\{5, 5\}$, and therefore their sum is 10.

Chapter 6
Vectors and Complex Numbers

Abstract A vector is a geometric object that has magnitude (or length) and direction and can be added to other vectors or multiplied by a scalar. Vectors play an important role in many branches of science, such as physics or engineering. They allow us to represent the position of objects in two, three or more dimensions and to store complex data having multiple components, such as velocity, temperature, pressure, etc. In this chapter we define vectors and their basic operations, such as addition, multiplication by a scalar, dot and cross products, together with some direct applications. The complex numbers in algebraic (vectors having one real and one imaginary component in the Argand diagram) and polar (vectors having radius and argument) form of are then presented, along with some of key operations and results.

Keywords Vector operations · Dot and cross product · Complex numbers · Algebraic and polar form · Operations with complex numbers

6.1 Brief Theoretical Background

In this section are presented notions about vectors and complex numbers.

Vectors

A geometric has **direction** as well as **magnitude**.
A vector is an ordered collection of n elements (components).

$\vec{a}$, $\underline{a}$, $\overrightarrow{AB}$, $\mathbf{a}$, $\hat{a}$

$$\mathbf{a} = (a_1, a_2, a_3), \quad \mathbf{a} = (a_x, a_y, a_z), \quad \mathbf{a} = [a_x, a_y, a_z]$$

O. Bagdasar, *Concise Computer Mathematics*, SpringerBriefs in Computer Science,
DOI: 10.1007/978-3-319-01751-8_6, © The Author(s) 2013

$$\mathbf{a} = \begin{bmatrix} a_1 \\ a_2 \\ a_3 \end{bmatrix} \text{(column)}, \quad \|\mathbf{a}\| \text{(length of vector } \mathbf{a})$$

$$A(a_1, a_2, a_3), B(b_1, b_2, b_3) \Rightarrow \overrightarrow{AB} = (b_1 - a_1, b_2 - a_2, b_3 - a_3).$$

Properties and Operations

- **Equality**: Two vectors are equal if they have same modulus and direction.
- **Zero vector**: $\mathbf{0} = (0, 0, 0) \in \mathbb{R}^3$ (3D) or $\mathbf{0} = (0, 0) \in \mathbb{R}^2$ (2D)
- **Unit vectors**: $\|\mathbf{a}\|$. Sometimes denoted by $\hat{a}$.
- **Addition**:
  - $\overrightarrow{AB} + \overrightarrow{BC} = \overrightarrow{AC}$ (triangle rule for geometric vectors)
  - Let $\mathbf{a} = (a_1, a_2, a_3)$ and $\mathbf{b} = (b_1, b_2, b_3)$ be two vectors. Their sum is

$$\mathbf{a} + \mathbf{b} = (a_1 + b_1, a_2 + b_2, a_3 + b_3).$$

- **Multiplication by a scalar**: Let $\mathbf{a} = (a_1, a_2, a_3)$ be a vector and $\lambda \in \mathbb{R}$. Then

$$\lambda\mathbf{a} = \lambda(a_1, a_2, a_3) = (\lambda a_1, \lambda a_2, \lambda a_3).$$

- **Subtraction**: $\mathbf{a} - \mathbf{b} = \mathbf{a} + (-1)\mathbf{b}$.
- **Polygon addition rule**: Let A, B, C, D be points in plane (space). Then

$$\overrightarrow{AB} + \overrightarrow{BC} + \overrightarrow{CD} + \overrightarrow{DA} = \mathbf{0}. \quad \text{(same is true for a polygon } A_1, \ldots, A_n)$$

Centre of Mass

The system of points $A_1, \ldots, A_n$ has the centre of mass M given by

$$\overrightarrow{OM} = \frac{\overrightarrow{OA_1} + \overrightarrow{OA_2} + \cdots + \overrightarrow{OA_n}}{n}.$$

If the points $A_1, \ldots, A_n$ of the system have masses $m_1, \ldots, m_n$ then the centre of mass M is given by

$$\overrightarrow{OM} = \frac{m_1\overrightarrow{OA_1} + m_2\overrightarrow{OA_2} + \cdots + m_n\overrightarrow{OA_n}}{m_1 + m_2 + \cdots + m_n}.$$

Dot (Scalar) Product

Let $\mathbf{a} = (a_1, a_2, a_3)$ and $\mathbf{b} = (b_1, b_2, b_3)$ be two vectors. The dot product is

$$\mathbf{a} \cdot \mathbf{b} = a_1 \cdot b_1 + a_2 \cdot b_2 + a_3 \cdot b_3$$
$$= \|a\|\|b\| \cos \theta$$

Applications of the Scalar Product

- The **modulus** of vector $\mathbf{a} = (a_1, a_2, a_3)$ is obtained from the formula

$$\|\mathbf{a}\| = \sqrt{\mathbf{a} \cdot \mathbf{a}} = \sqrt{a_1^2 + a_2^2 + a_3^2}$$

- The **angle** θ between vectors $\mathbf{a}$ and $\mathbf{b}$ is

$$\cos \theta = \frac{\mathbf{a} \cdot \mathbf{b}}{\|a\|\|b\|}$$

- Vectors $\mathbf{a}$ and $\mathbf{b}$ are **perpendicular** if and only if

$$\theta = \pi/2 \Leftrightarrow \cos \theta = 0 \Leftrightarrow \mathbf{a} \cdot \mathbf{b} = 0$$

- **Unit vectors** for $\mathbf{a}$ (same direction, modulus one):

$$\mathbf{a} = \|\mathbf{a}\|\hat{a}$$

Cross Product (Vector Product)

Let $\mathbf{a} = (a_1, a_2, a_3)$, $\mathbf{b} = (b_1, b_2, b_3)$ be vectors. The cross (vector) product is

$$\mathbf{a} \times \mathbf{b} = \|a\|\|b\| \sin \theta \, \hat{n} \quad (\hat{n} \text{ is the normal vector}).$$

The **normal vector on a, b** is the unit vector perpendicular on both $\mathbf{a}$ and $\mathbf{b}$.
Standard base: $\mathbf{i} = (1, 0, 0)$; $\mathbf{j} = (0, 1, 0)$; $\mathbf{k} = (0, 0, 1)$.
For these vectors we have

$$\mathbf{i} \times \mathbf{j} = \mathbf{k}; \quad \mathbf{j} \times \mathbf{i} = -\mathbf{k}$$
$$\mathbf{j} \times \mathbf{k} = \mathbf{i}; \quad \mathbf{k} \times \mathbf{j} = -\mathbf{i}$$
$$\mathbf{k} \times \mathbf{i} = \mathbf{j}; \quad \mathbf{i} \times \mathbf{k} = -\mathbf{j}$$
$$\mathbf{i} \times \mathbf{i} = \mathbf{j} \times \mathbf{j} = \mathbf{k} \times \mathbf{k} = \mathbf{0}$$

Vector product in coordinate notation: For coordinate vectors we have

$$\mathbf{a} = (a_1, a_2, a_3) = a_1\mathbf{i} + a_2\mathbf{j} + a_3\mathbf{k}.$$
$$\mathbf{b} = (b_1, b_2, b_3) = b_1\mathbf{i} + b_2\mathbf{j} + b_3\mathbf{k}.$$

the vector product is defined as

$$\mathbf{a} \times \mathbf{b} = (a_1\mathbf{i} + a_2\mathbf{j} + a_3\mathbf{k}) \times (b_1\mathbf{i} + b_2\mathbf{j} + b_3\mathbf{k})$$
$$= (a_2b_3 - a_3b_2)\mathbf{i} - (a_1b_3 - a_3b_1)\mathbf{j} + (a_1b_2 - a_2b_1)\mathbf{k}$$

Applications of the Cross (Vector) Product

- The **area** of a triangle between vectors $\mathbf{a} = (a_1, a_2, a_3)$ and $\mathbf{b} = (b_1, b_2, b_3)$ is obtained from the formula

$$S = \|\mathbf{a} \times \mathbf{b}\|$$

- The **angle** θ between vectors $\mathbf{a}$ and $\mathbf{b}$ is

$$\sin\theta = \frac{\|\mathbf{a} \times \mathbf{b}\|}{\|a\|\|b\|}$$

- Vectors $\mathbf{a}$ and $\mathbf{b}$ are **parallel** if and only if

$$\theta = 0 \Leftrightarrow \sin\theta = 0 \Leftrightarrow \mathbf{a} \times \mathbf{b} = 0$$

- Find a vector perpendicular to a plane. Take points $A, B, C \in \mathbb{R}^3$:

$$\overrightarrow{AB} \times \overrightarrow{AC} \perp \overrightarrow{AB}$$
$$\overrightarrow{AB} \times \overrightarrow{AC} \perp \overrightarrow{AC}$$

Fig. 6.1 Algebraic and polar representation of a complex number

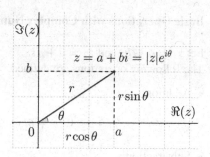

Complex Numbers

A complex number can be expressed in the form $a + bi$, where a and b are real numbers and $i = \sqrt{-1}$ is the imaginary number representing the solution of the equation $x^2 + 1 = 0$ (the notation j is used in Engineering) (Fig. 6.1).

Notation

$$z = a + bi, \quad a, b \in \mathbb{R} \qquad \text{(algebraic form)}$$
$$z = re^{i\theta} = r(\cos\theta + i\sin\theta) \quad r, \theta \in \mathbb{R} \qquad \text{(polar form)}$$
$$a = \Re(z) = r\cos\theta \qquad \text{(real part)}$$
$$b = \Im(z) = r\sin\theta \qquad \text{(imaginary part)}$$
$$r = |z| = \sqrt{a^2 + b^2} \qquad \text{(modulus)}$$
$$\theta \in [0, 2\pi) \qquad \text{(argument)}$$

Modulus Property

$$|z|^2 = z\bar{z} = (a + bi)(a - bi) = a^2 + b^2.$$

The Complex Conjugate

The conjugate of $z = a + bi$ is the number $\bar{z} = a - bi$.

Properties of the Conjugate

$$\overline{z_1 + z_2} = \overline{z_1} + \overline{z_2}$$
$$\overline{z_1 z_2} = \overline{z_1}\,\overline{z_2}$$
$$\overline{z_1/z_2} = \overline{z_1}/\overline{z_2}$$
$$\Re(z) = \frac{1}{2}(z + \bar{z}),$$
$$\Im(z) = \frac{1}{2i}(z - \bar{z})$$

Basic Operations with Complex Numbers

Let $z_1 = a + bi$ and $z_2 = c + di$.
The following operations are defined:

- Addition:

$$z_1 + z_2 = (a + bi) + (c + di) = (a + c) + (b + d)i$$

- Substraction:

$$z_1 - z_2 = (a + bi) - (c + di) = (a - c) + (b - d)i$$

- Multiplication (remember that $i^2 = -1$)

$$\begin{aligned}
z_1 z_2 &= (a + bi)(c + di) \\
&= ac + adi + bci + bdi^2 \\
&= (ac - bd) + (ad + bc)i
\end{aligned}$$

- Division (remember that $i^2 = -1$ and $z\bar{z} = |z|^2$)

$$\begin{aligned}
\frac{z_1}{z_2} &= \frac{a + bi}{c + di} = \frac{(a + bi)(c - di)}{(c + di)(c - di)} \\
&= \frac{ac + bd}{c^2 + d^2} + \frac{bc - ad}{c^2 + d^2}i
\end{aligned}$$

Example: Let $z_1 = 3 + 2i$ and $z_2 = 1 - 4i$. The following operations give

(1) $z_1 + z_2 = 4 - 2i$;
(2) $z_1 - z_2 = 2 + 6i$;
(3) $z_1 z_2 = 11 - 10i$;
(4) $|z_1| = \sqrt{3^2 + 2^2} = \sqrt{13}$;
(5) $z_1 / z_2 = -\dfrac{5}{17} + \dfrac{14}{17}i$.

De Moivre's formula

For the natural number $n \geq 2$ and $z = re^{i\theta} = (\cos\theta + i\sin\theta)$ we have

$$z^n = r^n e^{in\theta} = r^n(\cos n\theta + i \sin n\theta)$$

Remark: This formula is useful for computing powers of complex numbers.

6.2 Essential Questions

E 6.1. Decide which of the following are vector quantities: Velocity, mass, acceleration, weight, area, temperature, force, potential energy, volume.

E 6.2. In a system of vectors , a force of 10 newtons is represented by a line 5cm long. Find the magnitude of the force vector represented in the system by a line 2cm long.

E 6.3. (i) Find the magnitude of the following vectors:

$$\mathbf{a} = -2\mathbf{i} + 3\mathbf{j} + 4\mathbf{k}; \quad \mathbf{b} = \mathbf{i} + 2\mathbf{j} - 4\mathbf{k}; \quad \mathbf{c} = 3\mathbf{i} - 2\mathbf{j} + \mathbf{k}; \quad \mathbf{d} = 2\mathbf{j} + 3\mathbf{k}.$$

(ii) Find unit vectors parallel to the following vectors

$$\mathbf{u} = \mathbf{i} - 2\mathbf{j} + 4\mathbf{k}; \quad \mathbf{v} = 0.5\mathbf{i} + 2\mathbf{j} - \mathbf{k}; \quad \mathbf{w} = \mathbf{i} - 2\mathbf{j};$$

(iii) Find $x \in \mathbb{R}$ such that $\mathbf{c} = (x, 1, 1)$ is perpendicular on $\mathbf{u}$, $\mathbf{v}$ and $\mathbf{w}$.

E 6.4. (i) Evaluate the dot products of the following pairs of vectors:

$$(a) \quad \mathbf{p} = 2\mathbf{i} - 3\mathbf{j} + \mathbf{k}, \quad \mathbf{q} = 3\mathbf{i} + 4\mathbf{j} - 3\mathbf{k}$$
$$(b) \quad \mathbf{m} = 2\mathbf{i} - 3\mathbf{j} + 5\mathbf{k}, \quad \mathbf{l} = 3\mathbf{i} + 2\mathbf{j} - 2\mathbf{k}$$
$$(c) \quad \mathbf{v} = \mathbf{i} + 2\mathbf{j} + 2\mathbf{k}, \quad \mathbf{u} = -2\mathbf{i} - \mathbf{j} + 2\mathbf{k}$$

(ii) Use the definition of the dot product and your answers to part (i) to determine the angles between the pairs of vectors in part (i).

(iii) For each pair in (i), find a vector perpendicular on both vectors.

E 6.5. Evaluate the cross products of the following vectors:

$$(a) \quad \mathbf{p} = 2\mathbf{i} - \mathbf{j} + \mathbf{k}, \quad \mathbf{q} = 3\mathbf{i} + 4\mathbf{j} - \mathbf{k}$$
$$(b) \quad \mathbf{a} = \mathbf{i} + 2\mathbf{j} - 3\mathbf{k}, \quad \mathbf{b} = -2\mathbf{i} + 4\mathbf{j} + \mathbf{k}$$
$$(c) \quad \mathbf{r} = \mathbf{i} + 2\mathbf{j} - 3\mathbf{k}, \quad \mathbf{s} = 2\mathbf{i} + \mathbf{j} + 3\mathbf{k}.$$

E 6.6. Let $z_1 = 2 + 3i$ and $z_2 = 3 - 4i$. Compute the following

(1) $z_1 + z_2$; (2) $z_1 - z_2$; (3) $z_1 z_2$; (4) $|z_1|$; (5) z_1/z_2;

(6) $z_1 - \bar{z}_2$; (7) $z_1 \bar{z}_2$; (8) $z_2 \bar{z}_2$; (9) z_1^3; (10) z_1^4.

6.3 Supplementary Questions

S 6.1. Let ABCD be a quadrilateral. Find the single vectors equivalent to
(i) $\overrightarrow{AB} + \overrightarrow{BC}$; (ii) $\overrightarrow{AB} + \overrightarrow{BC} + \overrightarrow{CD}$; (iii) $\overrightarrow{AD} + \overrightarrow{DC}$; (iv) $\overrightarrow{BC} + \overrightarrow{CD}$; (v) $\overrightarrow{AB} + \overrightarrow{DA}$.

S 6.2. O, A, B, C, D are five points such that $\overrightarrow{OA} = \mathbf{a}$, $\overrightarrow{OB} = \mathbf{b}$, $\overrightarrow{OC} = \mathbf{a} + 2\mathbf{b}$ and $\overrightarrow{OD} = 2\mathbf{a} - \mathbf{b}$. Express $\overrightarrow{AB}$, $\overrightarrow{BC}$, $\overrightarrow{CD}$, $\overrightarrow{AC}$ and $\overrightarrow{BD}$ in terms of $\mathbf{a}$ and $\mathbf{b}$.

S 6.3. Let ABCD be a quadrilateral whose vertices are A(1,1), B(7,3), C(10,12), D(8,2). Find the vectors: (i) $\overrightarrow{AB}$; (ii) $\overrightarrow{AC}$; (iii) $\overrightarrow{AD}$; (iv) $\overrightarrow{BD}$;

S 6.4. For the quadrilateral ABCD in **S 6.3.**, check whether the diagonals $\overrightarrow{AC}$ and $\overrightarrow{BD}$ are perpendicular. If not, what is the angle between them ?

S 6.5. Let ABCD be a quadrilateral of vertices A(2,1), B(5,5), C(9,12), D(8,2).

  (i) Find the centre of mass G for the quadrilateral ABCD.

 (ii) Find the centres of mass G_A, G_B, G_C, G_D corresponding to each of the triangles BCD, ACD, ABD, ABC.

(iii) Find the centre of mass G_0 for the quadrilateral $G_A G_B G_C G_D$. What is the relation between G_0 and G?

(iv) Let $n \geq 4$, and let $a_1, \ldots, a_n$ and $b_1, \ldots, b_n$ be real numbers. Formulate statements similar to (i)–(iii) for the polygon $A_1(a_1, b_1), \ldots, A_n(a_n, b_n)$.

S 6.6. Answer same questions as in **S 6.5.** (i)–(iii), when masses 1, 2, 3 and 4 kg are attached to A, B, C and D, respectively.

S 6.7. Compute the following expressions:
(i) $(1 + i) + (3 - 4i)$; (ii) $\frac{1+i}{1-i}$; (iii) i^{2012}; (iv) i^{2013}; (v) i^{2014}; (vi) $(1 + i)^{2013}$.

S 6.8. Solve in $\mathbb{C}$ the following equations:

  (i) $z^2 + 4z + 2 = 0$;
 (ii) $z^2 + 2z + 4 = 0$;
(iii) $z^2 + 2z + 5 = 0$.
(iv) $iz^2 + (1 - 4i)z + 4i - 1 = 0$.

S 6.9. Let $z = 3 + 2i$. Find the complex number $u = a + bi$ s.t. $u^2 = z$.

S 6.10. Given that z is not real and $|z| = 1$, prove that the number $w = \frac{z-1}{z+1}$ is a pure imaginary number.

6.4 Problem Answers

Essential Problems

E 6.1. Vectors: velocity, acceleration, force.

E 6.2. 4 newtons.

E 6.3. (i) $\|\mathbf{a}\| = \sqrt{29}$; $\|\mathbf{b}\| = \sqrt{21}$; $\|\mathbf{c}\| = \sqrt{18}$; $\|\mathbf{d}\| = \sqrt{13}$;

(ii) $\hat{\mathbf{u}} = \frac{1}{\sqrt{21}}(\mathbf{i} - 2\mathbf{j} + 4\mathbf{k})$; $\hat{\mathbf{v}} = \frac{2}{\sqrt{21}}(0.5\mathbf{i} + 2\mathbf{j} - \mathbf{k})$; $\hat{\mathbf{w}} = \frac{1}{\sqrt{5}}(\mathbf{i} - 2\mathbf{j})$;

(iii) $\mathbf{c} \perp \mathbf{u}$: $x = -2$; $\mathbf{c} \perp \mathbf{v}$: $x = -2$; $\mathbf{c} \perp \mathbf{w}$: x=2.

E 6.4. (i) (a) $\mathbf{p} \cdot \mathbf{q} = -9$; (b) $\mathbf{m} \cdot \mathbf{l} = -10$; (c) $\mathbf{v} \cdot \mathbf{u} = 0$.

(ii) (a) $\cos \theta_{\mathbf{p},\mathbf{q}} = \frac{-9}{\sqrt{14}\sqrt{34}}$; (b) $\cos \theta_{\mathbf{m},\mathbf{l}} = \frac{-10}{\sqrt{38}\sqrt{17}}$; (c) $\cos \theta_{\mathbf{v},\mathbf{u}} = 0$.

(iii) In each case we have to find a vector $\mathbf{a} = (x, y, z)$ s.t.
(a) $\mathbf{a} \cdot \mathbf{p} = \mathbf{a} \cdot \mathbf{q} = 0$. Solution: $\mathbf{a} = (t, 9/5t, 17/10t)$, $t \in \mathbb{R}$.
(b) $\mathbf{a} \cdot \mathbf{m} = \mathbf{a} \cdot \mathbf{l} = 0$. Solution: $\mathbf{a} = (t, -7/2t, -16/13t)$, $t \in \mathbb{R}$.
(c) $\mathbf{a} \cdot \mathbf{v} = \mathbf{a} \cdot \mathbf{u} = 0$. Solution: $\mathbf{a} = (t, -t, 1/2t)$, $t \in \mathbb{R}$.

E 6.5. (a) $\mathbf{p} \times \mathbf{q} = -3\mathbf{i} + 5\mathbf{j} + 11\mathbf{k}$; (b) $\mathbf{a} \times \mathbf{b} = 14\mathbf{i} + 5\mathbf{j} + 8\mathbf{k}$; (c) $\mathbf{r} \times \mathbf{s} = 9\mathbf{i} - 9\mathbf{j} - 3\mathbf{k}$.

E 6.6. (1) $5 - i$; (2) $-1 + 7i$; (3) $18 + i$; (4) $\sqrt{13}$; (5) $(-6 + 17i)/25$; (6) $-1 - i$; (7) $-6 + 17i$; (8) 25; (9) $-46 + 9i$; (10) $119 - 12i$.

Supplementary Problems

S 6.1. (i) $\overrightarrow{AC}$; (ii) $\overrightarrow{AD}$; (iii) $\overrightarrow{AC}$; (iv) $\overrightarrow{BD}$; (v) $\overrightarrow{DB}$.

S 6.2. $\overrightarrow{AB} = \mathbf{b} - \mathbf{a}$; $\overrightarrow{BC} = \mathbf{a} + \mathbf{b}$; $\overrightarrow{CD} = \mathbf{a} - 3\mathbf{b}$; $\overrightarrow{AC} = 2\mathbf{b}$; $\overrightarrow{BD} = 2\mathbf{a} - 2\mathbf{b}$.

S 6.3. (i) $\overrightarrow{AB} = (6, 2)$; (ii) $\overrightarrow{AC} = (9, 11)$; (iii) $\overrightarrow{AD} = (7, 1)$; (iv) $\overrightarrow{BD} = (1, -1)$.

S 6.4. The two diagonals are perpendicular if the dot product is zero, i.e. $\overrightarrow{AC} \cdot \overrightarrow{BD} = 0$.

However, in this case the dot product is $\overrightarrow{AC} \cdot \overrightarrow{BD} = 9 \cdot 1 + 11 \cdot (-1) = -2 \neq 0$, which shows that the two vectors are not perpendicular. The cosine of the angle can obtained from the formula

$$\cos \theta_{\overrightarrow{AC},\overrightarrow{BD}} = \frac{-2}{\sqrt{202}\sqrt{2}} = \frac{-1}{\sqrt{101}}.$$

S 6.5. (i) Using the formula for the centre of mass one obtains

$$\overrightarrow{OG} = \frac{\overrightarrow{OA} + \overrightarrow{OB} + \overrightarrow{OC} + \overrightarrow{OD}}{4},$$

therefore $G(6, 5)$.
(ii) The centres are $G_A = (\frac{22}{3}, \frac{19}{3})$, $G_B = (\frac{19}{3}, 5)$, $G_C = (5, \frac{8}{3})$, $G_D = (\frac{16}{3}, 6)$.

(iii) One obtains

$$G_0 = \left(\frac{\frac{22}{3} + \frac{19}{3} + 5 + \frac{16}{3}}{4}, \frac{\frac{19}{3} + 5 + \frac{8}{3} + 6}{4} \right) = (6, 5) = G.$$

(iv) Let $n \geq 4$ be a natural number and let the polygon $A_1(a_1, b_1), \ldots, A_n(a_n, b_n)$ have the center of mass G. For the polygons A^i, $i = 1, \ldots, n$ (obtained by deleting the i-th vertex from the original polygon) having centre of mass G_i the following property holds: the centre of mass G_0 of the points $G_1, \ldots, G_n$ coincides with the center of mass G.

S 6.6. In this case the weighted formula for the center of mass yields

$$G_w = \frac{1(2, 1) + 2(5, 5) + 3(9, 12) + 4(8, 2)}{4}) = \left(\frac{71}{4}, \frac{55}{4} \right).$$

S 6.7. (i) $5 - 3i$; (ii) i; (iii) 1; (iv) i; (v) -1; (vi) $1 + i = \sqrt{2} e^{i\pi/4}$ therefore

$$(1 + i)^{2013} = \sqrt{2}^{2013} e^{i \frac{2013\pi}{4}} = \sqrt{2}^{2013} e^{i(126 \cdot 2\pi + \frac{5\pi}{4})}$$

$$= \sqrt{2}^{2013} (\cos \frac{5\pi}{4} + i \sin \frac{5\pi}{4}) = \sqrt{2}^{2013}(-1/2 - i/2).$$

S 6.8. Applying the normal quadratic formula one obtains the solutions

(i) $z_{1,2} = \frac{-4 \pm \sqrt{4^2 - 4 \cdot 2}}{2 \cdot 1} = -2 \pm \sqrt{2} \in \mathbb{R}$ (both solutions are real).

(ii) $z_{1,2} = \frac{-2 \pm \sqrt{2^2 - 4 \cdot 4}}{2 \cdot 1} = -1 \pm \sqrt{3} i \in \mathbb{C} \setminus \mathbb{R}$ (both solutions are complex).

(iii) $z_{1,2} = \frac{-2 \pm \sqrt{2^2 - 4 \cdot 5}}{2 \cdot 1} = -1 \pm 2i \in \mathbb{C} \setminus \mathbb{R}$ (both solutions are complex).

(iv) $z_{1,2} = \frac{-(1-4i) \pm \sqrt{(1-4i)^2 - 4i \cdot (4i - 1)}}{2 \cdot i} = \frac{1}{2} \left((4 + i) \pm \sqrt{4i - 1} \right).$

S 6.9. One needs to solve $(a + bi)^2 = a^2 - b^2 + 2abi = 3 + 2i$ which is equivalent to $a^2 - b^2 = 3$ and $ab = 1$ which gives $a^2 - 1/a^2 = 3$. Substituting $x = a^2$ the problem reduces to solving the quadratic $x^2 - 1 - 3x = 0$ with the solutions $x_{1,2} = \frac{1}{2}(3 \pm \sqrt{13})$. As $x = a^2$ the only solutions are $a_{1,2} = \pm \frac{1}{2}(3 + \sqrt{13})$ and $b_{1,2} = 1/a_{1,2}$.

S 6.10. Let $z = a + bi$. By amplification with $\bar{z} + 1$ one obtains

$$w = \frac{z - 1}{z + 1} = \frac{(\bar{z} + 1)(z - 1)}{(\bar{z} + 1)(z + 1)} = \frac{\bar{z}z - \bar{z} + z - 1}{|z + 1|^2} = \frac{2bi}{|z + 1|^2} = \frac{2b}{|z + 1|^2} i.$$

Chapter 7
Matrices and Applications

Abstract A matrix is a rectangular array of numbers, symbols, or expressions (called entries or elements) arranged in rows and columns. Matrices are direct generalisations of vectors and play a key role in many mathematical areas such as linear algebra or computer graphics (where they are used to define linear transformations). In this chapter we define matrices and illustrate their properties through examples. We then present some basic matrix operations such as addition and multiplication, as well as the determinant and inverse of square matrices. Finally, matrices are used for solving systems of linear equations. These results prepare the introduction of matrix based linear transformations in computer graphics, discussed in the next chapter.

Keywords Matrix operations · Determinant · Inverse · Systems of equations

7.1 Brief Theoretical Background

A matrix is an array of numbers called **elements**. Horizontal components are called **rows** (or lines) while the vertical components are called **columns**. A matrix with m rows and n columns is of **dimension** (size, order) $m \times n$. Two matrices are **equal** if they have same dimension and elements.

Example: The following matrix has **2 rows** and **4 columns**

$$\begin{pmatrix} 2 & 0 & 1 & 3 \\ 1 & 3 & 1 & 2 \end{pmatrix}.$$

Special Types of Matrices (Orderwise)

- row matrix: $\begin{pmatrix} 1 & 0 & 1 \end{pmatrix}$ or $\begin{bmatrix} 1 & 0 & 2 \end{bmatrix}$

O. Bagdasar, *Concise Computer Mathematics*, SpringerBriefs in Computer Science,
DOI: 10.1007/978-3-319-01751-8_7, © The Author(s) 2013

- column matrix: $\begin{pmatrix} 1 \\ 0 \\ 2 \end{pmatrix}$ or $\begin{bmatrix} 1 \\ 0 \\ 2 \end{bmatrix}$

- square matrix (i.e. $m = n$): $\begin{pmatrix} 2 & 0 \\ 1 & 3 \end{pmatrix}$ or $\begin{bmatrix} 2 & 0 \\ 1 & 3 \end{bmatrix}$

Matrix Notation

- Matrices are usually denoted by capital letters.

$$\mathbf{A} = \begin{bmatrix} 1 & 0 & 2 \end{bmatrix}$$

- a matrix of order $m \times n$ is denoted by

$$\mathbf{A} = \begin{bmatrix} a_{11} & \cdots & a_{1j} & \cdots & a_{1n} \\ \vdots & \ddots & \vdots & \ddots & \vdots \\ a_{i1} & \cdots & a_{ij} & \cdots & a_{in} \\ \vdots & \ddots & \vdots & \ddots & \vdots \\ a_{m1} & \cdots & a_{mj} & \cdots & a_{mn} \end{bmatrix}$$

- a_{ij} denotes the element in row i and column j.

- A matrix of order $m \times n$ in X is a also a **function**

$$\mathbf{A} : \{1, \ldots, m\} \times \{1, \ldots, n\} \to X, \quad \mathbf{A}(i, j) = a_{ij}.$$

Special Matrices

- **The zero matrix:** all its elements are zero. The zero 2×2 matrix is

$$\underline{\mathbf{0}} = \begin{bmatrix} 0 & 0 \\ 0 & 0 \end{bmatrix}$$

- **Diagonal matrix:** Square, elements outside the main diagonal are zero.

$$\mathbf{A} = \begin{bmatrix} -1 & 0 \\ 0 & 2 \end{bmatrix}, \quad \mathbf{B} = \begin{bmatrix} -1 & 0 & 0 \\ 0 & 3 & 0 \\ 0 & 0 & 7 \end{bmatrix}, \ldots$$

- **The identity matrix:** Diagonal matrix with 1 on the main diagonal.

$$\mathbf{I}_2 = \begin{bmatrix} 1 & 0 \\ 0 & 1 \end{bmatrix}, \quad \mathbf{I}_3 = \begin{bmatrix} 1 & 0 & 0 \\ 0 & 1 & 0 \\ 0 & 0 & 1 \end{bmatrix}, \ldots$$

- **Lower/Upper triangular matrix:** All Elements below/above main diagonal are zero.

$$\mathbf{L} = \begin{bmatrix} 3 & 0 \\ 1 & 2 \end{bmatrix}, \quad \mathbf{U} = \begin{bmatrix} 2 & 2 & 4 \\ 0 & 4 & 3 \\ 0 & 0 & 1 \end{bmatrix}, \ldots$$

Matrix Operations

- **Matrix transpose:** changes columns with rows.
  First column becomes first row, second column becomes second row...

$$\text{If } \mathbf{A} = \begin{bmatrix} a & b \\ c & d \end{bmatrix}, \text{ then } \mathbf{A}^T = \begin{bmatrix} a & c \\ b & d \end{bmatrix}$$

Remark 1: The transpose of an $m \times n$ matrix is an $n \times m$ matrix.

Remark 2: A matrix for which $\mathbf{A} = \mathbf{A}^T$ is called **symmetric**.

- **Addition and subtraction:**

  - element by element operation
  - matrices need to be of the same order (dimension, size)
  - addition and subtraction are done element by element.

- **Multiplication by a scalar:** Let $\mathbf{A}$ be a matrix and $\lambda \in \mathbb{R}$ a scalar (number). The matrix $\lambda\mathbf{A}$ is obtained by multiplying each element of $\mathbf{A}$ by λ.

$$\lambda\mathbf{A} = \lambda \begin{bmatrix} a & b \\ c & d \end{bmatrix} = \begin{bmatrix} \lambda a & \lambda b \\ \lambda c & \lambda d \end{bmatrix}$$

- **Multiplication of two matrices:** the rule of **row-column multiplication**. (Fig. 7.1) Let $\mathbf{A}$, $\mathbf{B}$ be matrices of sizes $m \times n$ and $n \times p$ respectively. The element c_{ij} in $\mathbf{C} = \mathbf{AB}$ is obtained by multiplying

  - the i-th row of $\mathbf{A}$ by
  - the j-th column of $\mathbf{B}$.

Example: $A(3 \times 2) \times B(2 \times 4) = C(3 \times 4)$.

- **Determinant** (2×2 matrix): The determinant of matrix $\mathbf{A} = \begin{bmatrix} a & b \\ c & d \end{bmatrix}$ is

$$|\mathbf{A}| = \begin{vmatrix} a & b \\ c & d \end{vmatrix} = ad - bc = \det A$$

Fig. 7.1 Matrix multiplication diagram

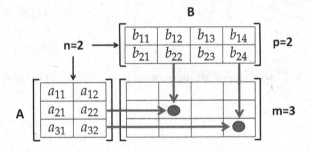

- **Inverse** (2×2 matrix): The inverse of matrix $\mathbf{A} = \begin{bmatrix} a & b \\ c & d \end{bmatrix}$ is

$$\mathbf{A}^{-1} = \frac{1}{|\mathbf{A}|} \begin{bmatrix} d & -b \\ -c & a \end{bmatrix}$$

Remark 3: The inverse only exists when $|\mathbf{A}| \neq 0$!

Remark 4: A matrix for which $\mathbf{A}^{-1} = \mathbf{A}^T$ is called **orthogonal**.

Remark 5: Determinants exist for any square matrix of higher order.

Solution of a Linear System of Equations

Example: Consider the system of linear equations

$$2x + 3y = 13$$
$$3x + 4y = 18$$

This can be written in matrix form as

$$\begin{pmatrix} 2 & 3 \\ 3 & 4 \end{pmatrix} \begin{bmatrix} x \\ y \end{bmatrix} = \begin{bmatrix} 13 \\ 18 \end{bmatrix}$$

The solution is

$$\begin{bmatrix} x \\ y \end{bmatrix} = \underline{x} = \mathbf{A}^{-1}\underline{b} = (-1) \begin{pmatrix} 4 & -3 \\ -3 & 2 \end{pmatrix} \begin{bmatrix} 13 \\ 18 \end{bmatrix} = \begin{bmatrix} 2 \\ 3 \end{bmatrix}$$

Remark 6: Any linear systems of equations has a matrix representation.

7.2 Essential Questions

E 7.1. State the dimension of each of the following matrices:

(a) $A = \begin{bmatrix} 1 & 2 & 3 \\ 2 & 1 & 2 \end{bmatrix}$, (b) $B = \begin{bmatrix} 2 & 4 \\ 2 & 3 \\ 3 & 1 \end{bmatrix}$, (c) $C = \begin{bmatrix} 1 & 2 \end{bmatrix}$, (d) $D = \begin{bmatrix} 1 \\ 2 \\ 3 \end{bmatrix}$.

E 7.2. Indicate whether the following statements are true or false, for the matrices below

(a) $A = \begin{bmatrix} 5 & 3 & 1 \\ 2 & 1 & 3 \end{bmatrix}$, (b) $B = \begin{bmatrix} 5 & 3 & 1 \end{bmatrix}$,

(c) $C = \begin{bmatrix} 10/2 & 6/2 & 1 \end{bmatrix}$, (d) $D = \begin{bmatrix} 1 & 2 \\ 2 & 3 \end{bmatrix}$, (e) $E = \begin{bmatrix} 3 & 2 \\ 2 & 1 \end{bmatrix}$.

(i) $D = A$; (ii) $D = E$; (iii) $B = C$; (iv) $A = B$; (v) $D - C = \underline{0}$.

E 7.3. Let A, B and C be the matrices defined below

$$A = \begin{bmatrix} 5 & 3 & 1 \\ 2 & 1 & 3 \end{bmatrix}, \quad B = \begin{bmatrix} 1 & 2 & 3 \\ 4 & 3 & 2 \end{bmatrix}, \quad C = \begin{bmatrix} 0 & 1 & 1 \\ 2 & 3 & 2 \end{bmatrix}.$$

Compute each of the following

(i) $A + C$; (ii) $A - B$; (iii) $2A$; (iv) $-3C$; (v) $A - 2B + 4C$; (vi) B^T; (vii) $B + C^T$.

E 7.4. Form the product of the following matrices, stating whether the result is a column or row vector:

(i) $\begin{bmatrix} 5 & 3 \end{bmatrix} \begin{bmatrix} 2 & 3 \\ 1 & 2 \end{bmatrix}$, (ii) $\begin{bmatrix} 5 & 3 & 1 \end{bmatrix} \begin{bmatrix} 2 & 3 & 1 \\ 1 & 2 & 4 \\ 3 & 1 & 2 \end{bmatrix}$,

(iii) $\begin{bmatrix} 5 & 3 \\ 2 & 3 \end{bmatrix} \begin{bmatrix} 2 \\ 3 \end{bmatrix}$, (iv) $\begin{bmatrix} 5 \\ 3 \\ 1 \end{bmatrix} \begin{bmatrix} 1 & 2 & 4 \end{bmatrix}$.

E 7.5. Find the determinant of the matrix $\begin{bmatrix} 5 & 3 \\ 4 & 2 \end{bmatrix}$, and find its inverse. Use the result to solve the pair of simultaneous equations

$$\begin{cases} 5x + 3y = 19 \\ 4x + 2y = 14 \end{cases}$$

E 7.6. Let the matrices A, B, C be

$$A = \begin{bmatrix} 5 & 3 & 1 \\ 2 & 1 & 3 \\ 3 & 1 & 2 \end{bmatrix}, \quad B = \begin{bmatrix} 1 & 2 & 3 \\ 4 & 3 & 2 \\ 2 & 1 & 3 \end{bmatrix}, \quad C = \begin{bmatrix} 0 & 1 & 1 \\ 2 & 3 & 2 \\ 5 & 2 & 1 \end{bmatrix}.$$

Compute (i) AB; (ii) AC; (iii) CA. Show that $AC \neq CA$, therefore matrix multiplication is generally not commutative.

7.3 Supplementary Questions

S 7.1. Compute (i) $A + B$; (ii) $A^T + B^T$; (iii) $(A + B)^T$ for the matrices

$$\text{Let } A = \begin{bmatrix} 5 & 3 \\ 2 & 1 \\ 3 & 1 \end{bmatrix} \text{ and } B = \begin{bmatrix} 1 & 3 \\ 4 & 2 \\ 2 & 3 \end{bmatrix}.$$

What is the connection between the results for parts (ii) and (iii) above?

S 7.2. Consider the matrices defined below

$$C = \begin{bmatrix} 2 & 3 \\ 4 & 2 \\ 3 & 1 \end{bmatrix} \text{ and } D = \begin{bmatrix} 4 & 2 \\ 3 & 1 \\ 2 & 3 \end{bmatrix}.$$

Compute (i) $C + D$; (ii) $C - D$; (iii) $(C - D)^T$; (iv) $C + C^T$; (v) $(C^T)^T$ for

S 7.3. Compute AI and IA for the matrices

$$A = \begin{bmatrix} 5 & 3 \\ 2 & 1 \end{bmatrix} \text{ and } I = \begin{bmatrix} 1 & 0 \\ 0 & 1 \end{bmatrix}.$$

What does this suggest about the multiplication by the identity matrix?

S 7.4. Evaluate the result of the operations (i) A^T; (ii) AA^T; (iii) $A^T A$; (iv) AA and (v) $A^T A^T$ for the matrix $A = \begin{bmatrix} 5 & 3 & 1 \\ 2 & 1 & 3 \end{bmatrix}$. In general, if A is an $m \times n$ matrix, what can you say about the size of matrices AA^T and $A^T A$?

S 7.5. Given the following matrices

$$A = \begin{bmatrix} 5 & 3 & 1 \end{bmatrix}, \quad B = \begin{bmatrix} 1 \\ 2 \\ 3 \end{bmatrix}, \quad C = \begin{bmatrix} 0 & 1 \\ 2 & 3 \end{bmatrix}, \quad D = \begin{bmatrix} 2 & 3 & 2 \\ 5 & 2 & 1 \end{bmatrix}.$$

State the size of the following products and compute the ones that exist.

(i) **AB**; (ii) **BA**; (iii) **CD**; (iv) **DB**; (v) **DC**; (vi) **BD**; (vii) $\mathbf{A}^2$; (viii) $\mathbf{C}^2$.

S 7.6. Evaluate the determinants of the following matrices:

$$A = \begin{bmatrix} 5 & 3 \end{bmatrix}, \quad B = \begin{bmatrix} 1 & 2 \\ 3 & 5 \end{bmatrix}, \quad C = \begin{bmatrix} 2 & 6 \\ 1 & 3 \end{bmatrix}, \quad D = \begin{bmatrix} 4 & 2 \\ 3 & 5 \end{bmatrix}, \quad E = \begin{bmatrix} 2 & 3 & 2 \\ 5 & 2 & 1 \\ 1 & 4 & 3 \end{bmatrix}.$$

Find the inverses of the 2 × 2 matrices **B**, **C** and **D**.

S 7.7. Solve the following systems of linear equations using the inverse matrix method.

(i) $\begin{cases} 3x - y = 4 \\ 2x - y = -11 \end{cases}$ (ii) $\begin{cases} 4x - 3y = 6 \\ 3x - 4y = -9 \end{cases}$ (iii) $\begin{cases} -2x + y = -4 \\ x + 2y = 2 \end{cases}$

S 7.8. State the reason why the inverse matrix method cannot be applied to the system

$$\begin{cases} 4x + 6y = 12 \\ 2x + 3y = 14 \end{cases}$$

S 7.9. Write the following system in matrix form

$$\begin{cases} 4x + 6y + 3z = 28 \\ 2x + 3y + 2z = 15 \\ 3x + 2y + 5z = 19 \end{cases}$$

7.4 Problem Answers

Essential Problems

E 7.1. (a) 2 × 3; (b) 3 × 2; (c) 1 × 2; (d) 3 × 1.

E 7.2. (a) F; (b) F; (c) T; (d) F; (e) F.

E 7.3. (i) $A + C = \begin{bmatrix} 5 & 4 & 2 \\ 4 & 4 & 5 \end{bmatrix}$; (ii) $A - B = \begin{bmatrix} 4 & 1 & -2 \\ -2 & -2 & 1 \end{bmatrix}$; (iii) $2A = \begin{bmatrix} 10 & 6 & 2 \\ 4 & 2 & 6 \end{bmatrix}$;

(iv) $-3C = \begin{bmatrix} 0 & -3 & -3 \\ -6 & -9 & -6 \end{bmatrix}$; (v) $A - 2B + 4C = \begin{bmatrix} 3 & 3 & -1 \\ 2 & 7 & 7 \end{bmatrix}$; (vi) $B^T = \begin{bmatrix} 1 & 4 \\ 2 & 3 \\ 3 & 2 \end{bmatrix}$;

(vii) operation not defined.

E 7.4. (i) row matrix: $\begin{bmatrix} 13 & 21 \end{bmatrix}$; (ii) row matrix: $\begin{bmatrix} 16 & 22 & 19 \end{bmatrix}$;

(iii) column matrix: $\begin{bmatrix} 16 \\ 13 \end{bmatrix}$;   (iv) square matrix: $\begin{bmatrix} 5 & 10 & 20 \\ 3 & 6 & 12 \\ 1 & 2 & 4 \end{bmatrix}$.

E 7.5. $\det \mathbf{A} = -2$; $\mathbf{A}^{-1} = \begin{bmatrix} -1 & 1.5 \\ 2 & -2.5 \end{bmatrix}$; $x = 2$; $y = 3$.

E 7.6. (i) $\mathbf{AB} = \begin{bmatrix} 19 & 20 & 24 \\ 12 & 10 & 17 \\ 11 & 11 & 17 \end{bmatrix}$;   (ii) $\mathbf{AC} = \begin{bmatrix} 11 & 16 & 12 \\ 17 & 11 & 7 \\ 12 & 10 & 7 \end{bmatrix}$;   (iii) $\mathbf{CA} = \begin{bmatrix} 5 & 2 & 5 \\ 22 & 11 & 15 \\ 32 & 18 & 13 \end{bmatrix}$.

Supplementary problems

S 7.1. The results are the following

$$(i)\ \mathbf{A} + \mathbf{B} = \begin{bmatrix} 6 & 6 \\ 6 & 3 \\ 5 & 4 \end{bmatrix};\quad (ii)\ \mathbf{A}^T + \mathbf{B}^T = \begin{bmatrix} 6 & 6 & 5 \\ 6 & 3 & 4 \end{bmatrix};\quad (iii)\ (\mathbf{A} + \mathbf{B})^T = \begin{bmatrix} 6 & 6 & 5 \\ 6 & 3 & 4 \end{bmatrix}.$$

One can see that $\mathbf{A}^T + \mathbf{B}^T = (\mathbf{A} + \mathbf{B})^T$, therefore the transpose can be applied at any stage when adding matrices.

S 7.2. The results are

$$(i)\ \mathbf{C} + \mathbf{D} = \begin{bmatrix} 6 & 5 \\ 7 & 3 \\ 5 & 4 \end{bmatrix};\quad (ii)\ \mathbf{C} - \mathbf{D} = \begin{bmatrix} -2 & 1 \\ 1 & 1 \\ 1 & -2 \end{bmatrix};\quad (iii)\ (\mathbf{C} - \mathbf{D})^T = \begin{bmatrix} -2 & 1 & 1 \\ 1 & 1 & -2 \end{bmatrix};$$

(iv) The addition is not compatible; (v) $(\mathbf{C}^T)^T = \mathbf{C}$.

S 7.3. By doing the multiplication one can check that $\mathbf{AI} = \mathbf{IA} = \mathbf{A}$, which is valid for any square matrix $\mathbf{A}$ and the corresponding identity matrix.

S 7.4. The matrix operations give

$$(i)\ \mathbf{A}^T = \begin{bmatrix} 5 & 2 \\ 3 & 1 \\ 1 & 3 \end{bmatrix};\quad (ii)\ \mathbf{AA}^T = \begin{bmatrix} 35 & 16 \\ 16 & 14 \end{bmatrix};\quad (iii)\ \mathbf{A}^T\mathbf{A} = \begin{bmatrix} 29 & 17 & 11 \\ 17 & 10 & 6 \\ 11 & 6 & 10 \end{bmatrix}.$$

The operations defined in (iv) and (v) are not compatible. For a matrix $\mathbf{A}$ of size $m \times n$, the size of matrix $\mathbf{AA}^T$ is $m \times m$ while the size of matrix $\mathbf{A}^T\mathbf{A}$ is $n \times n$. Also, one may note that these two matrices are symmetric.

S 7.5. The answers are

$$(i)\ \mathbf{AB} = 14;\quad (ii)\ \mathbf{BA} = \begin{bmatrix} 5 & 3 & 1 \\ 10 & 6 & 2 \\ 15 & 9 & 3 \end{bmatrix};\quad (iii)\ \mathbf{CD} = \begin{bmatrix} 5 & 2 & 1 \\ 19 & 12 & 7 \end{bmatrix};\quad (iv)\ \mathbf{DB} = \begin{bmatrix} 14 \\ 12 \end{bmatrix}.$$

The operations (v) **DC**, (vi) **BD** and (vii) $\mathbf{A}^2$ are not compatible.
The answer for (viii) is $\mathbf{CD} = \begin{bmatrix} 2 & 3 \\ 6 & 11 \end{bmatrix}$.

S 7.6. Clearly, det $\mathbf{A} = |\mathbf{A}|$ does not exist. For the other matrices we have

$$|\mathbf{B}| = -1; \quad |\mathbf{C}| = 0; \quad |\mathbf{D}| = 14; \quad |\mathbf{E}| = -2.$$

Matrix **C** is not invertible, while for the other ones we have

$$\mathbf{B}^{-1} = \begin{bmatrix} -5 & 2 \\ 3 & -1 \end{bmatrix}, \quad \mathbf{D}^{-1} = \frac{1}{14} \begin{bmatrix} 5 & -2 \\ -3 & 4 \end{bmatrix}.$$

S 7.7. The solutions can be obtained from the formulae

(i) $\begin{bmatrix} x \\ y \end{bmatrix} = \underline{x} = \mathbf{A}^{-1}\underline{b} = (-1) \begin{pmatrix} -1 & 1 \\ -2 & 3 \end{pmatrix} \begin{bmatrix} 4 \\ -11 \end{bmatrix} = \begin{bmatrix} 15 \\ 41 \end{bmatrix}$

(ii) $\begin{bmatrix} x \\ y \end{bmatrix} = \underline{x} = \mathbf{A}^{-1}\underline{b} = \frac{1}{-7} \begin{pmatrix} -4 & 3 \\ -3 & 4 \end{pmatrix} \begin{bmatrix} 6 \\ -9 \end{bmatrix} = \frac{1}{7} \begin{bmatrix} 51 \\ 54 \end{bmatrix}$

(iii) $\begin{bmatrix} x \\ y \end{bmatrix} = \underline{x} = \mathbf{A}^{-1}\underline{b} = \frac{1}{-5} \begin{pmatrix} 2 & -1 \\ -1 & -2 \end{pmatrix} \begin{bmatrix} -4 \\ 2 \end{bmatrix} = \begin{bmatrix} 2 \\ 0 \end{bmatrix}$

S 7.8. The inverse matrix method may not be applied to the system because the determinant of the matrix $\begin{pmatrix} 4 & 6 \\ 2 & 3 \end{pmatrix}$ is zero.

S 7.9. The system can be written as

$$\begin{cases} 4x + 6y + 3z = 28 \\ 2x + 3y + 2z = 15 \\ 3x + 2y + 5z = 19 \end{cases} \iff \begin{pmatrix} 4 & 6 & 3 \\ 2 & 3 & 2 \\ 3 & 2 & 5 \end{pmatrix} \begin{pmatrix} x \\ y \\ z \end{pmatrix} = \begin{pmatrix} 28 \\ 15 \\ 19 \end{pmatrix}.$$

Chapter 8
Matrix Applications in Computer Graphics

Abstract Column matrices can be used to represent points in $2D$ or $3D$, while matrices of dimension $2 \times n$ and $3 \times n$ can be used to represent sets of points in $2D$ or $3D$. Matrices allow arbitrary linear transformations to be represented in a consistent format ($T(\mathbf{x}) = \mathbf{Ax}$ for some $2 \times n$ (or $3 \times n$) matrix $\mathbf{A}$, called the transformation matrix of T), suitable for computation. This format allows transformations to be conveniently combined with each other by multiplying their matrices. In this chapter we first use matrices to represent points, lines and polygons. We then discuss in detail some linear transformations such as translation, scaling, rotation, reflections and shearing in 2D, and examine how transformations can be concatenated using matrix multiplication.

Keywords Points · Linear transformations · Rotations · Reflections · Shearing.

8.1 Brief Theoretical Background

Here we use matrices to represent linear transformations in 2D.

Matrix Representation

- **Points:** represented by a column matrix (position vector)

$$X = \begin{bmatrix} x \\ y \end{bmatrix}.$$

- **Lines:** defined by two points and represented by a 2×2 matrix.
  The matrix $\begin{bmatrix} 3 & -1 \\ 2 & 4 \end{bmatrix}$ represents a line through points $\begin{bmatrix} 3 \\ 2 \end{bmatrix}$ and $\begin{bmatrix} -1 \\ 4 \end{bmatrix}$.

O. Bagdasar, *Concise Computer Mathematics*, SpringerBriefs in Computer Science, DOI: 10.1007/978-3-319-01751-8_8, © The Author(s) 2013

- **Polygons:** Represented by a $2 \times n$ matrix.

  The matrix $\begin{bmatrix} 3 & -1 & 1 \\ 2 & 4 & 1 \end{bmatrix}$ represents the triangle $\begin{bmatrix} 3 \\ 2 \end{bmatrix}$, $\begin{bmatrix} -1 \\ 4 \end{bmatrix}$ and $\begin{bmatrix} 1 \\ 1 \end{bmatrix}$.

Matrix Transformations of Points

Consider the point X and a transformation matrix T given by

$$X = \begin{bmatrix} x \\ y \end{bmatrix}, \quad T = \begin{bmatrix} a & b \\ c & d \end{bmatrix}.$$

Matrix T pre-multiplies point X, generating another point X^*

$$TX = \begin{bmatrix} a & b \\ c & d \end{bmatrix} \begin{bmatrix} x \\ y \end{bmatrix} = \begin{bmatrix} ax + by \\ cx + dy \end{bmatrix} = \begin{bmatrix} x^* \\ y^* \end{bmatrix} = X^*$$

Example: The point $X = \begin{bmatrix} -1 \\ 2 \end{bmatrix}$ is transformed by the matrix $T = \begin{bmatrix} 1 & 2 \\ 3 & 4 \end{bmatrix}$.

$$X^* = \begin{bmatrix} x^* \\ y^* \end{bmatrix} = TX = \begin{bmatrix} 1 & 2 \\ 3 & 4 \end{bmatrix} \begin{bmatrix} -1 \\ 2 \end{bmatrix} = \begin{bmatrix} 1(-1) + 2(2) \\ 3(-1) + 4(2) \end{bmatrix} = \begin{bmatrix} 3 \\ 5 \end{bmatrix}$$

Transformation of Lines

- Take two points X, Y and a transformation T
- Apply the transformation T to both X, Y (they determine a line XY)
- The transformed points X^*, Y^* determine a **new line** X^*Y^*

Remark: Matrix transformations change lines into lines.

Example: A straight line joins the points $(2, 2)$ and $(4, 5)$.

Find the new line produced by the transformation matrix $T = \begin{bmatrix} 4 & 2 \\ 3 & 1 \end{bmatrix}$.

Solution: The new line is determined by the two points representing the columns of the matrix

$$\begin{bmatrix} 4 & 2 \\ 3 & 1 \end{bmatrix} \begin{bmatrix} 2 & 4 \\ 2 & 5 \end{bmatrix} = \begin{bmatrix} 12 & 26 \\ 8 & 17 \end{bmatrix}.$$

Linear Transformations in 2D

Consider the point X, vector V and a transformation matrix T given by

$$X = \begin{bmatrix} x \\ y \end{bmatrix}, \quad V = \begin{bmatrix} u \\ v \end{bmatrix}, \quad T = \begin{bmatrix} a & b \\ c & d \end{bmatrix}.$$

The following linear transformations can be defined.

Translation: The translation of point X by vector V is defined by

$$X^* = \begin{bmatrix} x^* \\ y^* \end{bmatrix} = X + V = \begin{bmatrix} x+u \\ y+v \end{bmatrix}.$$

Scaling: Scales horizontal and vertical coordinates.

$$T = \begin{bmatrix} \alpha & 0 \\ 0 & \beta \end{bmatrix}; \quad X^* = \begin{bmatrix} x^* \\ y^* \end{bmatrix} = TX = \begin{bmatrix} \alpha & 0 \\ 0 & \beta \end{bmatrix}\begin{bmatrix} x \\ y \end{bmatrix} = \begin{bmatrix} \alpha x \\ \beta y \end{bmatrix}.$$

Rotation: The line from $(0, 0)$ to the point (x, y), is rotated by an angle θ. By convention **anticlockwise movement is considered positive**. We say that (x, y) has rotated about the origin to (x^*, y^*).

$$T = \begin{bmatrix} \cos\theta & -\sin\theta \\ \sin\theta & \cos\theta \end{bmatrix}.$$

Reflections

- in the X axis: **changes the sign of the y co-ordinate**

$$T = \begin{bmatrix} 1 & 0 \\ 0 & -1 \end{bmatrix}; \quad X^* = TX = \begin{bmatrix} 1 & 0 \\ 0 & -1 \end{bmatrix}\begin{bmatrix} x \\ y \end{bmatrix} = \begin{bmatrix} x \\ -y \end{bmatrix}$$

- in the Y axis: **changes the sign of the x co-ordinate**

$$T = \begin{bmatrix} -1 & 0 \\ 0 & 1 \end{bmatrix}; \quad X^* = TX = \begin{bmatrix} -1 & 0 \\ 0 & 1 \end{bmatrix}\begin{bmatrix} x \\ y \end{bmatrix} = \begin{bmatrix} -x \\ y \end{bmatrix}$$

- in the $y = x$ line: **interchanges the x and y co-ordinates**.

$$T = \begin{bmatrix} 0 & 1 \\ 1 & 0 \end{bmatrix}; \quad X^* = TX = \begin{bmatrix} 0 & 1 \\ 1 & 0 \end{bmatrix}\begin{bmatrix} x \\ y \end{bmatrix} = \begin{bmatrix} y \\ x \end{bmatrix}$$

- in the $y = -x$ line: **interchanges x and y co-ordinates and changes signs**.

$$T = \begin{bmatrix} 0 & -1 \\ -1 & 0 \end{bmatrix}; \quad X^* = TX = \begin{bmatrix} 0 & -1 \\ -1 & 0 \end{bmatrix}\begin{bmatrix} x \\ y \end{bmatrix} = \begin{bmatrix} -y \\ -x \end{bmatrix}$$

Shearing: A shear is a **distortion**. The transformation matrix

$$T = \begin{bmatrix} a & b \\ c & d \end{bmatrix}$$

The 'off-diagonal' elements b and c determine the kind of shear produced.

- b produces x-direction shear,
- c produces y-direction shear.

Combinations of Linear Transformations

Translations, rotations, scalings or shears are very often combined to generate more complex transformations. In the language of matrices, this translates into computing the product of the matrices corresponding to each individual transformation.

If the point X is subject to two consecutive transformations T_1 and T_2, then the composed transformation $T = T_2 T_1$ generates the new point

$$X^* = T_2(T_1 X) = (T_2 T_1)X = TX.$$

Example: The point $(3, 2)$ is transformed by the scaling $\begin{bmatrix} 3 & 0 \\ 0 & 2 \end{bmatrix}$, then reflected in Y axis and finally rotated by $45°$. Find the coordinates of the new point.

Solution: Using the notations

$$T_1 = \begin{bmatrix} 3 & 0 \\ 0 & 2 \end{bmatrix}, \ T_2 = \begin{bmatrix} -1 & 0 \\ 0 & 1 \end{bmatrix}, \ T_3 = \begin{bmatrix} \cos 45° & -\sin 45° \\ \sin 45° & \cos 45° \end{bmatrix} = \frac{1}{\sqrt{2}} \begin{bmatrix} \cos 1 & -1 \\ \sin 1 & 1 \end{bmatrix},$$

the coordinate of the new point is given by

$$X^* = T_3 T_2 T_1 X = \frac{1}{\sqrt{2}} \begin{bmatrix} 1 & -1 \\ 1 & 1 \end{bmatrix} \begin{bmatrix} -1 & 0 \\ 0 & 1 \end{bmatrix} \begin{bmatrix} 3 & 0 \\ 0 & 2 \end{bmatrix} \begin{bmatrix} 3 \\ 2 \end{bmatrix} = \frac{1}{\sqrt{2}} \begin{bmatrix} -13 \\ -5 \end{bmatrix}.$$

Remark: Matrices of size 3×3 can represent 3D linear transformations.

8.2 Essential Questions

E 8.1. A straight line joins the points $(3, 2)$ and $(4, 5)$.
Describe the transformed line produced by the matrix $T = \begin{bmatrix} 4 & 2 \\ 3 & -1 \end{bmatrix}$.

E 8.2. Write down a transformation matrix which

(a) scales vertical lines by a factor of 3 leaving horizontal lines unchanged.
(b) scales horizontal lines by a factor of 2 and vertical lines by a factor of 5.

E 8.3. The point (3, 1) is rotated anticlockwise about the origin through an angle of 45°. Calculate the position of the new point.

E 8.4. Write down the following matrices.

(a) Rotation about the origin, anti-clockwise through $\pi/6$ radians.
(b) Rotation about the origin, clockwise through $\pi/6$.
(c) Use the appropriate matrix to find the new position of the point (2, 3) after it has been rotated anti-clockwise through $\pi/6$ about the origin.
(d) Calculate the product of the two matrices found in (a) and (b). Explain your answer.

E 8.5. Consider the 2D unit square having coordinates (0, 0), (1, 0), (0, 1), (1, 1). Determine the new co-ordinates of the figure after the transformations:

(1) Translation of vector (1, 2);
(2) horizontal scaling of factor 2;
(3) vertical scaling of factor 3;
(4) mixed scaling of factors 3 and 4;
(5) Rotation of angle $\pi/6$;
(6) Rotation of angle 90°;
(7) Reflection in the X axis;
(8) Reflection in the Y axis;
(9) Reflection in the $y = x$ line;
(10) Reflection in the $y = -x$ line.

E 8.6. Point (2, 3) is reflected in Y axis, then reflected in $y = x$, and then rotated by 90°. Find the co-ordinates of the final point.

8.3 Supplementary Questions

S 8.1. Show that any rotation matrix R has the property $R^T = R^{-1}$ (i.e. the transpose of a rotation matrix is equal to its inverse).

S 8.2. Let R be the rotation matrix about the origin through angle θ.

(a) State the rotation matrix S for a **clockwise** rotation through an angle θ.
(b) Calculate R^{-1} using the result of problem **S 8.1.** above.
(c) What can be concluded from parts (a) and (b) above.

S 8.3. Write down transformation matrices which will:

(a) scale vertical lines by a factor of 5 and leave horizontal lines unchanged.
(b) scale horizontal lines by a factor of 4 and vertical lines by a factor 0.25.
(c) Use the above matrices to find the new co-ordinates of the points (2, 1) and (−2, −1), performing each transformation separately.

S 8.4. Point A(2, −1) is reflected in **X** axis, then rotated by 90° and finally reflected in $y = -x$. Find the co-ordinates of the final point.

S 8.5. Find the coordinates of the unit square defined in **E 8.5.**, subject to the shear transformations given by the matrices:

(a) $T = \begin{bmatrix} 1 & 3 \\ 0 & 1 \end{bmatrix}$; (b) $T = \begin{bmatrix} 1 & 0 \\ 2 & 1 \end{bmatrix}$; (c) $T = \begin{bmatrix} 1 & 2 \\ 3 & 1 \end{bmatrix}$; (d) $T = \begin{bmatrix} 2 & 3 \\ 4 & 5 \end{bmatrix}$.

8.4 Problem Answers

Essential Problems

E.8.1 The first line is $\begin{bmatrix} 3 & 4 \\ 2 & 5 \end{bmatrix}$. The transformed line is $\begin{bmatrix} 4 & 2 \\ 3 & -1 \end{bmatrix}\begin{bmatrix} 3 & 4 \\ 2 & 5 \end{bmatrix} = \begin{bmatrix} 16 & 26 \\ 7 & 7 \end{bmatrix}$.

E 8.2. (a) $T = \begin{bmatrix} 3 & 0 \\ 0 & 1 \end{bmatrix}$; (b) $T = \begin{bmatrix} 5 & 0 \\ 0 & 2 \end{bmatrix}$.

E 8.3. The new point is given by: $T\begin{bmatrix} 1 \\ 3 \end{bmatrix} = \begin{bmatrix} \cos 45° & -\sin 45° \\ \sin 45° & \cos 45° \end{bmatrix}\begin{bmatrix} 3 \\ 1 \end{bmatrix} = \begin{bmatrix} \sqrt{2} \\ 2\sqrt{2} \end{bmatrix}$.

E 8.4. Using the formulae for the rotation matrix we obtain

(a) $R = \begin{bmatrix} \cos \pi/6 & -\sin \pi/6 \\ \sin \pi/6 & \cos \pi/6 \end{bmatrix} = \begin{bmatrix} \sqrt{3}/2 & -1/2 \\ 1/2 & \sqrt{3}/2 \end{bmatrix}$.

(b) $S = \begin{bmatrix} \cos(-\pi/6) & -\sin(-\pi/6) \\ \sin(-\pi/6) & \cos(-\pi/6) \end{bmatrix} = \begin{bmatrix} \sqrt{3}/2 & 1/2 \\ -1/2 & \sqrt{3}/2 \end{bmatrix}$.

(c) $R\begin{bmatrix} 2 \\ 3 \end{bmatrix} = \begin{bmatrix} \cos \pi/6 & -\sin \pi/6 \\ \sin \pi/6 & \cos \pi/6 \end{bmatrix}\begin{bmatrix} 2 \\ 3 \end{bmatrix} = \begin{bmatrix} \sqrt{3}/2 & -1/2 \\ 1/2 & \sqrt{3}/2 \end{bmatrix}\begin{bmatrix} 2 \\ 3 \end{bmatrix} = \begin{bmatrix} 0.2321 \\ 3.5981 \end{bmatrix}$.

(d) The product is the identity matrix, as the two transformations cancel each other.

E 8.5. First, the unit square is represented by the matrix $\begin{bmatrix} 0 & 1 & 0 & 1 \\ 0 & 0 & 1 & 1 \end{bmatrix}$.

(1) $\begin{bmatrix} 0 & 1 & 0 & 1 \\ 0 & 0 & 1 & 1 \end{bmatrix} + \begin{bmatrix} 1 & 1 & 1 & 1 \\ 2 & 2 & 2 & 2 \end{bmatrix} = \begin{bmatrix} 1 & 2 & 1 & 2 \\ 2 & 2 & 3 & 3 \end{bmatrix}$.

(2) $T = \begin{bmatrix} 2 & 0 \\ 0 & 1 \end{bmatrix}$; $T\begin{bmatrix} 0 & 1 & 0 & 1 \\ 0 & 0 & 1 & 1 \end{bmatrix} = \begin{bmatrix} 0 & 2 & 0 & 2 \\ 0 & 0 & 1 & 1 \end{bmatrix}$.

(3) $T = \begin{bmatrix} 1 & 0 \\ 0 & 3 \end{bmatrix}$; $T\begin{bmatrix} 0 & 1 & 0 & 1 \\ 0 & 0 & 1 & 1 \end{bmatrix} = \begin{bmatrix} 0 & 1 & 0 & 1 \\ 0 & 0 & 3 & 3 \end{bmatrix}$.

(4) $T = \begin{bmatrix} 3 & 0 \\ 0 & 4 \end{bmatrix}$; $T\begin{bmatrix} 0 & 1 & 0 & 1 \\ 0 & 0 & 1 & 1 \end{bmatrix} = \begin{bmatrix} 0 & 3 & 0 & 3 \\ 0 & 0 & 4 & 4 \end{bmatrix}$.

(5) $T = \begin{bmatrix} \cos \pi/6 & -\sin \pi/6 \\ \sin \pi/6 & \cos \pi/6 \end{bmatrix}$;

(6) $T = \begin{bmatrix} 0 & -1 \\ 1 & 1 \end{bmatrix}$

(7) $T = \begin{bmatrix} 1 & 0 \\ 0 & -1 \end{bmatrix}$; $T\begin{bmatrix} 0 & 1 & 0 & 1 \\ 0 & 0 & 1 & 1 \end{bmatrix} = \begin{bmatrix} 0 & 1 & 0 & 1 \\ 0 & 0 & -1 & -1 \end{bmatrix}$.

(8) $T = \begin{bmatrix} -1 & 0 \\ 0 & 0 \end{bmatrix}$; $T\begin{bmatrix} 0 & 1 & 0 & 1 \\ 0 & 0 & 1 & 1 \end{bmatrix} = \begin{bmatrix} 0 & -1 & 0 & -1 \\ 0 & 0 & 1 & 1 \end{bmatrix}$.

(9) $T = \begin{bmatrix} 0 & 1 \\ 1 & 0 \end{bmatrix}$; $T\begin{bmatrix} 0 & 1 & 0 & 1 \\ 0 & 0 & 1 & 1 \end{bmatrix} = \begin{bmatrix} 0 & 0 & 1 & 1 \\ 0 & 1 & 0 & 1 \end{bmatrix}$.

(10) $T = \begin{bmatrix} 0 & -1 \\ -1 & 0 \end{bmatrix}$; $T\begin{bmatrix} 0 & 1 & 0 & 1 \\ 0 & 0 & 1 & 1 \end{bmatrix} = \begin{bmatrix} 0 & -1 & 0 & -1 \\ 0 & 0 & -1 & -1 \end{bmatrix}$.

E 8.6. The following matrix transformations are involved

$$T_1 = \begin{bmatrix} -1 & 0 \\ 0 & 1 \end{bmatrix}, \quad T_2 = \begin{bmatrix} 0 & 1 \\ 1 & 0 \end{bmatrix}; \quad T_3 = \begin{bmatrix} 0 & -1 \\ 1 & 0 \end{bmatrix}$$

The final transformation is

$$T = T_3 T_2 T_1 = \begin{bmatrix} -1 & 0 \\ 0 & 1 \end{bmatrix}\begin{bmatrix} 0 & 1 \\ 1 & 0 \end{bmatrix}\begin{bmatrix} 0 & -1 \\ 1 & 0 \end{bmatrix} = \begin{bmatrix} 0 & -1 \\ -1 & 0 \end{bmatrix}$$

The co-ordinates of the transformed point are $(-3, -2)$.

Supplementary Problems

S 8.1. The transformation matrix for a rotation of angle θ is $R(\theta) = \begin{bmatrix} \cos \theta & -\sin \theta \\ \sin \theta & \cos \theta \end{bmatrix}$, whose inverse is the rotation of angle $-\theta$, which satisfies

$$R(-\theta) = \begin{bmatrix} \cos(-\theta) & -\sin(-\theta) \\ \sin(-\theta) & \cos(-\theta) \end{bmatrix} = \begin{bmatrix} \cos \theta & \sin \theta \\ -\sin \theta & \cos \theta \end{bmatrix} = R(\theta)^T,$$

because $\sin(-\theta) = -\sin(\theta)$ and $\cos(-\theta) = \cos(\theta)$.

S 8.2. The answer can easily be obtained from **S 8.3.** and **E 8.4.**

S 8.3. (a) $T = \begin{bmatrix} 5 & 0 \\ 0 & 1 \end{bmatrix}$; (b) $T = \begin{bmatrix} 1/4 & 0 \\ 0 & 4 \end{bmatrix}$. (c) $(10,1),(-10,-1);(1/2,8),(-1/2,-8)$.

S 8.4. The following matrix transformations are involved

$$T_1 = \begin{bmatrix} 1 & 0 \\ 0 & -1 \end{bmatrix}, \quad T_2 = \begin{bmatrix} 0 & -1 \\ 1 & 0 \end{bmatrix}, \quad T_3 = \begin{bmatrix} 0 & -1 \\ -1 & 0 \end{bmatrix}$$

The final transformation is

$$T = T_3 T_2 T_1 = \begin{bmatrix} 0 & -1 \\ -1 & 0 \end{bmatrix}\begin{bmatrix} 0 & -1 \\ 1 & 0 \end{bmatrix}\begin{bmatrix} 1 & 0 \\ 0 & -1 \end{bmatrix} = \begin{bmatrix} -1 & 0 \\ 0 & -1 \end{bmatrix}$$

The co-ordinates of the transformed point are $(-2, 1)$.

S 8.5. (a) $\begin{bmatrix} 0 & 1 & 3 & 4 \\ 0 & 0 & 1 & 1 \end{bmatrix}$; (b) $\begin{bmatrix} 0 & 1 & 0 & 1 \\ 0 & 2 & 1 & 3 \end{bmatrix}$; (c) $\begin{bmatrix} 0 & 1 & 2 & 3 \\ 0 & 3 & 1 & 4 \end{bmatrix}$; (d) $\begin{bmatrix} 0 & 2 & 3 & 5 \\ 0 & 4 & 5 & 9 \end{bmatrix}$.

Chapter 9
Elements of Graph Theory

Abstract Graphs are mathematical structures used to model and visualize relations between certain objects. One of the first formulations of a graph theory problem was the famous Konigsberg Problem solved by Leonhard Euler. He proposed a model which reduced the problem to a schematic diagram and then concluded that the graph needed to satisfy some general conditions for the problem to be solved affirmatively. His studies highlighted the importance of understanding graphs and their properties, which later lead to the creation of Graph Theory as a distinct mathematical discipline. In this chapter we first introduce graphs through some illustrative examples and then describe the basic elements of graphs, as well some important graph properties. We then present special types of graphs like trees or networks, together with some specific techniques, problems and algorithms.

Keywords Graph elements · Trees · Networks · Adjacency matrix · Cost matrix.

9.1 Brief Theoretical Background

A graph is a mathematical structure used to model pairwise relations between objects from a certain collection. Before exploring specific properties, two illustrative examples are presented.

Example 1: The Konigsberg Bridge Problem (Eighteenth Century) In Fig. 9.1a are sketched seven bridges (lines) which connect four islands (nodes). The question is whether it possible to start on one island, walk over each of the seven bridges exactly once and then return to the starting point?

Solution: The answer is no, and was found by Euler in 1736. His idea was to represent the problem using 4 nodes (land masses) and 7 arcs (bridges). The graph should have had exactly zero or two nodes of odd degree for a solution to exist.

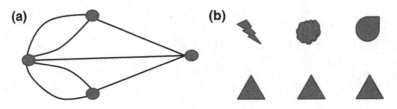

Fig. 9.1 a Konigsberg bridges diagram; **b** Utility problem diagram

Example 2: The utility problem (Fig. 9.1b)

 We have 3 houses and 3 utility companies - say, gas, electric, and water.
 Is it possible to connect each utility to each house without any crossovers ?

Solution: The answer to the puzzle posed in Fig. 9.1b is NO.

 The problem asks whether the complete bipartite graph $K_{3,3}$ is planar.

Basic elements of a graph G

- **vertices:** (nodes) $V(G)$: the set of points
- **edges:** (arcs) $E(G)$: lines connecting two vertices (directed or not)
- **loops:** edge that connects a vertex to itself
- **order** (degree) of a node: number of 'incident' edges.
- **trail:** sequence of arcs s.t. the end node of one arc is the start of the next.
- **path:** trail in which no node is passed through more than once
- **cycle:** path with an extra arc joining the final node to the initial node.

Example:

$$V(G) = \{1, 2, 3, 4, 5, 6\}.$$
$$E(G) = \{(1, 1), (1, 2), (3, 4), (4, 6), (4, 5), (5, 1), (5, 2)\}.$$
$$\text{ord}(3) = 2.$$
loop: $(1, 1)$
trail: $1 - 2 - 3 - 4 - 5 - 2 - 3$.
path: $1 - 2 - 3 - 4 - 5 - 6$.
cycle: $1 - 2 - 5 - 1$.

Special Graphs

- **directed:** Some edges have a direction.
- **connected:** It is possible to reach any node from any node
- **complete:** Simple, undirected, arcs between all nodes (Notation: K_n)
- **planar:** It can be drawn so that the arcs do not cross
- **simple:** A graph with no loops or multiple arcs
- **tree:** Simple graph with no cycles

- **spanning tree** (for a connected, undirected graph): subgraph that is a tree and connects all the vertices together.
- **network:** A graph with weighted arcs (could be distance, cost, time).

Adjacency matrix: For a graph G with n vertices this is a $n \times n$ matrix s.t.

- non-diagonal entry a_{ij}: number of edges from vertex i to vertex j
- diagonal entry a_{ii}: number of edges from vertex i to itself taken **once** for directed graphs or **twice** for undirected graphs

Example: For the graph depicted in the previous page we have

	1	2	3	4	5	6
1	2	1	0	0	1	0
2	1	0	1	0	1	0
3	0	1	0	1	0	0
4	0	0	1	0	1	1
5	1	1	0	1	0	0
6	0	0	0	1	0	0

Trees: classification and basic operations

- **sequence:** each node has 1 predecessor (or 0) and 1 successor (or 0)
- **binary tree:** each node has 1 predecessor (or 0) and 2 successors (or 1, 0)
- **general tree:** can have any number of successors

Standard operations on binary trees

The set of all trees with vertices in set X is denoted by $BTREE(X)$.
For a given tree $t \in BTREE(X)$ the following operations are defined

- **ROOT:** $BTREE(X) \rightarrow X$ (finds root)
- **LEFT:** $BTREE(X) \rightarrow BTREE(X)$ (finds left branch)
- **RIGHT:** $BTREE(X) \rightarrow BTREE(X)$ (finds right branch)
- **ISEMPTYTREE:** $BTREE(X) \rightarrow B$ (checks if tree is empty)

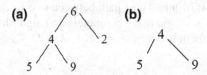

For tree t in (a) below, $ROOT(t) = 6$, $LEFT(t)$ is tree (b) while $RIGHT(t) = 2$.
For t and r trees in $BTREE(X)$ and $s \in X$ one can define the function:

- **MAKE:** $BTREE(X) \times X \times BTREE(X) \rightarrow BTREE(X)$; $(t, s, r) \mapsto u$
  This output u is a tree with left branch t, root s and right branch r.

Network problems

Networks can be represented using diagrams or a distance matrix.

Network   Distance matrix

	A	B	C	D	E	F	G
A	-	7	-	5	-	-	-
B	7	-	8	9	7	-	-
C	-	8	-	-	5	-	-
D	5	9	-	-	15	6	-
E	-	7	5	15	-	8	9
F	-	-	-	6	8	-	11
G	-	-	-	-	9	11	-

Kruskal's algorithm (minimum spanning tree)

- List the arcs in order of increasing weight
- Choose the arc with least weight
- Build a tree by working down the list choosing arcs provided they do **not** form a cycle when added to the arcs already chosen
- Stop when no more arcs can be chosen.

Example: The segments selected when using Kruskal's algorithm are

$AD = 5$, $CE = 5$, $DF = 6$, $AB = 7$, $BE = 7$ (not $BC = 8$, $FE = 8$, $BD = 9$) and $EG = 9$. Segments $FG = 11$ and $DE = 15$ do not contribute.

Prims's algorithm (minimum spanning tree)

- Choose a node
- Build a tree by choosing the minimum weight arc joining a node not yet chosen to one that has. Add this arc and the end node to the tree
- Repeat the tree building process until all the nodes have been chosen.

Example: The segments selected when using Prim's algorithm are $AD = 5$, $DF = 6$, $AB = 7$, $BE = 7$, $CE = 5$ and $EG = 9$.

Shortest path problem: Minimal cost path between two vertices in a graph. For large networks this is solved using computers and algorithms.
Example: The shortest path between A and G is $A - D - F - G$ of cost 22.

9.2 Essential Problems

E 9.1. Prove that the sum of degrees of vertices in any finite graph is even.

E 9.2. Draw the graphs represented by the following adjacency matrices:

E 9.3. (a) Write down the information in **E 9.2.** as a relation $R \subset X \times X$ where $X = \{a, b, c, d, e\}$ i.e. as a set of ordered pairs.

		a	b	c	d	e
	a	0	1	1	0	1
(i)	b	1	0	1	1	1
	c	1	1	0	0	0
	d	0	1	0	0	1
	e	1	1	0	1	0

		a	b	c	d	e
	a	0	1	0	0	1
(ii)	b	1	0	1	1	0
	c	0	1	1	1	0
	d	0	1	1	0	1
	e	1	0	0	1	1

		a	b	c	d	e
	a	0	1	0	0	1
(iii)	b	0	0	0	0	0
	c	0	0	0	0	0
	d	0	0	0	0	1
	e	0	0	0	0	0

(b) Which if any of the graphs in **E 9.2.** have the following properties: connected, complete, directed, undirected, contain a cycle, contain a loop, is a tree, contains a vertex which is even, contains a vertex which is odd?

E 9.4. In a group of people John likes Mary, Brian and Emma; Brian likes Mary and Sue; Mary likes John and Sue; Emma likes Mary, John and Brian; Sue likes nobody. Draw a graph showing who likes who.

E 9.5. Given the trees p and q from BTREE (BIRDS) defined below

Write down the following:

 (i) LEFT(RIGHT(p));
 (ii) ISEMPTYTREE(RIGHT(RIGHT(q)));
(iii) ROOT(LEFT(p));
(iv) MAKE(q, gull, q);
 (v) MAKE (LEFT(p), ROOT(q), RIGHT(p));
(vi) MAKE(LEFT(LEFT(q)), ROOT(p), MAKE(q, puffin, RIGHT(p))).

E 9.6. Given the tree u from BTREE(S) defined below:

Evaluate each of the following:

  (i) LEFT(RIGHT(LEFT(u)));
  (ii) RIGHT(LEFT(RIGHT(u)));
  (iii) ROOT(u) $+_S$ ROOT(LEFT(RIGHT(u))).

9.3 Supplementary Problems

S 9.1. Show that a tree with n vertices has exactly $n - 1$ edges.

S 9.2. Prove that a complete graph with n nodes (K_n) has $n(n - 1)/2$ edges.

S 9.3. Show that every simple graph has two vertices of the same degree.

S 9.4. Using standard functions of BTREE(X), describe the functions.

  (i) F that inputs a vertex and a binary tree, and replaces the root of the tree by the input vertex.
  (ii) G that inputs a binary tree and exchanges left and right subtrees. Give an example in each case to illustrate what the function does.

S 9.5. Use standard functions of BTREE(X) to give a formal description of two functions that input a binary tree and an item from X with the actions:

  (i) REP_L – replaces the root of the left subtree by the input item.
  (ii) REP_R – replaces the root of the right subtree by the input item.

S 9.6. Let a network be given by the distance matrix

	A	B	C	D	E	F	G
A	–	10	5	4	3	–	–
B	10	–	8	9	7	–	2
C	5	8	–	–	5	–	–
D	4	9	–	–	10	6	4
E	3	7	5	10	–	8	9
F	–	–	–	6	8	–	11
G	–	2	–	4	9	11	–

Find the minimum spanning tree using

  (a) Kruskal's algorithm (node and arc list, start with A).
  (b) Prim's algorithm (arc list).
  (c) Find the shortest path between A and G.

9.4 Problem Answers

Essential problems

E 9.1. Each arc contributes with 2 to the sum of degrees, therefore this sum should be twice the number of arcs.

E 9.2. Draw the vertices $\{a, b, c, d, e\}$ and directed segments from x to y if $(x, y) = 1$ in each of the adjacency matrices.

E 9.3. We write down the directed segments corresponding to each matrix

(i) $G_1 = \{(a, b), (a, c), (a, e), (b, a), (b, c), (b, d), (b, e), (c, a), (c, b), (d, b),$
$(d, e), (e, a), (e, b), (e, d)\}$;

(ii) $G_2 = \{(a, b), (a, e), (b, a), (b, c), (b, d), (c, b), (c, c), (c, d)(d, b), (d, c),$
$(d, e)(e, a), (e, d), (e, e)\}$;

(iii) $G_3 = \{(a, b), (a, e), (d, e)\}$.

E 9.4. The adjacency matrix is

	John	Mary	Brian	Emma	Sue
John	0	1	1	1	0
Mary	1	0	0	0	1
Brian	0	1	0	0	1
Emma	1	1	1	0	0
Sue	0	0	0	0	0

Drawing the graph is straightforward.

E 9.5. (i) swan;   (ii) False;   (iii) hawk;

 iv) Tree whose left branch is q, root is "gull" and right branch is q;

 v) Tree whose left branch is LEFT(p), root is "rook" and right branch is RIGHT(p);

 vi) Tree whose left branch is empty, root is "eagle" and right branch is another tree; this smaller tree has left branch q, root "puffin" and right branch RIGHT(p).

E 9.6. (i) empty;   (ii) empty;   (iii) "adamwitch".

Supplementary Problems

S 9.1. A tree with n vertices has to be connected, but also to have no cycles. We can imagine that initially there is just one node. To add another node to this one, one has to draw a segment. This step is repeated with the addition of any new node and in the end one obtains exactly $n - 1$ edges.

S 9.2. In a complete graph each vertex is connected to all the others. This means that in total we have $n(n - 1)$ edges. As using this formula each node was counted twice, the total number of edges has to be $n(n - 1)/2$ or $\binom{n}{2}$.

S 9.3. Let G be any finite simple graph with $n \neq 2$ vertices. The maximal degree of any vertex in G is less than equal to $n - 1$. Also, if our graph G is not connected, then the maximal degree is strictly less than $n - 1$.

Case 1: Assume that G is connected. We can not have a vertex of degree 0 in G, so the set of vertex degrees is a subset of $S = \{1, 2, , n - 1\}$. Since the graph G has n vertices, by pigeon-hole principle we can find two vertices of the same degree in G.

Case 2: Assume that G is not connected. G has no vertex of degree $n - 1$, so the set of vertex degrees is a subset of $S_0 = 0, 1, 2, , n - 2$. By pigeon-hole principle again, we can find two vertices of the same degree in G.

S 9.4. The functions are defined as:

(i) $F : BTREE(X) \times X \to BTREE(X)$

$$F(T, x) = MAKE\left(LEFT(T), x, RIGHT(T)\right)$$

(ii) $G : BTREE(X) \to BTREE(X)$

$$G(T) = MAKE\left(RIGHT(T), ROOT(T), LEFT(T)\right).$$

S 9.5. (i) The function $REP_L : BTREE(X) \times X \to BTREE(X)$ is defined as

$$REP_L(T, x) = MAKE\,(MAKE\,(LEFT(LEFT(T)), x, RIGHT(LEFT(T))),$$
$$ROOT(T), RIGHT(T))$$

(ii) The function $REP_R : BTREE(X) \times X \to BTREE(X)$ is defined as

$$REP_R(T, x) = MAKE\Big(LEFT(T), ROOT(T),$$
$$MAKE\Big(LEFT(RIGHT(T)), x, RIGHT(RIGHT(T))\Big)\Big)$$

S 9.6. (a) Nodes used by Kruskal's algorithm are $A - E - D - G - B - C - F$. Arcs: $AE = 3$, $AD = 4$, $DG = 4$, $BG = 2$, $AC = 5$ and $DF = 6$ with cost 24.

(b) The arcs used by Prim's algorithm are $BG = 2$, $AE = 3$, $AD = 4$, $DG = 4$, $AC = 5$ (or $CE = 5$) and $DF = 6$ for a total of 24.

(c) The shortest path between A and G is $A - D - G$ of length 8.

Chapter 10
Elements of Number Theory and Cryptography

Abstract Number theory is an important mathematical domain dedicated to the study of numbers and their properties. As discussed in Chap 1, the number systems $\mathbb{N}, \mathbb{Z}, \mathbb{Q}, \mathbb{C}$ emerged from the need of solving more complicated equations. However, numerous fundamental (some still unanswered) questions in mathematics refer to prime numbers and their properties (Goldbach conjecture, Rieman hypothesis). Cryptography studies techniques for a secure communication in the presence of adversaries which involves coding (encrypting) and decoding (decrypting). From the rudimentary tools used since antiquity (letter shifting, sticks), modern cryptography makes extensive use of mathematics, information theory, computational complexity, statistics, combinatorics, abstract algebra and number theory. In this chapter we present basic elements of number theory including prime numbers, divisibility, Euler's totient function and modulo arithmetic, which are used to describe the Caesar cypher and the RSA algorithm.

Keywords Prime factorisation · Least common multiple · Greatest common divisor · Totient function · Modulo arithmetic · Caesar cypher · RSA algorithm

10.1 Brief Theoretical Background

Prime Numbers and Divisibility

A number n is **prime** if his only positive divisors are 1 and itself.

Fundamental theorem of arithmetic (unique factorization theorem):
Every integer $n \geq 1$ is either prime or is the product of prime numbers.
Every integer $n > 1$ is represented **uniquely** as a product of prime powers

$$n = p_1^{a_1} p_2^{a_2} \cdots p_k^{a_k}, \quad \text{(canonical representation)}$$

where $p_1 < p_2 < ... < p_k$ are primes and a_i are positive integers.

O. Bagdasar, *Concise Computer Mathematics*, SpringerBriefs in Computer Science,
DOI: 10.1007/978-3-319-01751-8_10, © The Author(s) 2013

Greatest common divisor: $\gcd(a, b)$ or (a, b)

Example: What is the greatest common divisor of 54 and 24?
Solution: The divisors of 54 are: $1, 2, 3, 6, 9, 18, 27, 54$.
The divisors of 24 are: $1, 2, 3, 4, 6, 8, 12, 24$.
The common divisors of 54 and 24: $1, 2, 3, 6$.

The greatest of these is 6, written as $\gcd(54, 24) = 6$.

Least common multiple: l cm (a, b) or $[a, b]$

Example: What is the LCM of 4 and 6?
Solution: The multiples of 4 are: $4, 8, 12, 16, 20, 24, 28, 32, 36, \ldots$
The multiples of 6 are: $6, 12, 18, 24, 30, 36, 42, 48, 54, \ldots$
Common multiples of 4 and 6 are $12, 24, 36, 48, 60, 72, \ldots$

The least common multiple is 12.

Definition Numbers a and b are **relatively prime** if $\gcd(a, b) = 1$.

Properties of gcd(a, b) and l cm(a, b): If $a = p_1^{a_1} p_2^{a_2} \cdots p_k^{a_k}$, $\quad b = p_1^{b_1} p_2^{b_2} \cdots p_k^{b_k}$
where $p_1 < p_2 < \ldots < p_k$ are primes and a_i, b_i are non-negative integers.

The following properties hold:

$$a \cdot b = p_1^{a_1+b_1} \, p_2^{a_2+b_2} \, \cdots \, p_k^{a_k+b_k}$$
$$\gcd(a, b) = p_1^{\min(a_1,b_1)} \, p_2^{\min(a_2,b_2)} \, \cdots \, p_k^{\min(a_k,b_k)}$$
$$\mathrm{lcm}(a, b) = p_1^{\max(a_1,b_1)} \, p_2^{\max(a_2,b_2)} \, \cdots \, p_k^{\max(a_k,b_k)}$$
$$a \cdot b = \mathrm{lcm}(a, b) \cdot \gcd(a, b)$$

Euler's Totient Function $\varphi(n)$

The number of integers $1 \leq k \leq n$ relatively prime with n.

Example: The set of numbers prime with 8 are $1, 3, 5, 7$, therefore $\varphi(8) = 4$.
Formula properties: If p is prime, $\gcd(m, n) = 1$ and $k > 1$ then

$$\varphi(p) = p - 1$$
$$\varphi(mn) = \varphi(m)\varphi(n).$$

Formula (Euler): If the factorisation of n is $n = p_1^{a_1} p_2^{a_2} \cdots p_k^{a_k}$ then

$$\varphi(n) = n \left(1 - \frac{1}{p_1}\right) \left(1 - \frac{1}{p_2}\right) \cdots \left(1 - \frac{1}{p_k}\right).$$

Example: $\varphi(36) = \varphi\left(2^2 3^2\right) = 36 \left(1 - \frac{1}{2}\right) \left(1 - \frac{1}{3}\right) = 36 \cdot \frac{1}{2} \cdot \frac{2}{3} = 12$.

Modular Arithmetic

System of arithmetic for integers (clock arithmetic), where numbers "wrap around" upon reaching a certain value called **modulus**.

Definition For $n \in \mathbb{N}$, two integers a and b are called **congruent modulo** n:

$$a \equiv b \pmod{n},$$

if their difference $a - b$ is an integer multiple of n.

Modulo operations

If $a_1 \equiv b_1 \pmod{n}$ and $a_2 \equiv b_2 \pmod{n}$ we can define

- Addition: $a_1 + a_2 \equiv b_1 + b_2 \pmod{n}$

  - $2 + 1 \pmod{6} \equiv 3 \pmod{6}$
  - $2 + 4 \pmod{6} \equiv (2 + 4) \pmod{6} = 0 \pmod{6}$

- Substraction $a_1 - a_2 \equiv b_1 - b_2 \pmod{n}$

  - $2 - 1 \pmod{6} \equiv 1 \pmod{6}$
  - $2 - 4 \pmod{6} \equiv (2 - 4) \pmod{6} = 4 \pmod{6}$

- Multiplication $a_1 a_2 \equiv b_1 b_2 \pmod{n}$.

  - $2 \times 2 \pmod{6} \equiv 4 \pmod{6}$
  - $2 \times 3 \pmod{6} \equiv 6 \pmod{6} = 0 \pmod{6}$

- Inversion $a_1 / a_2 \equiv b_1 / b_2 \pmod{n}$ (only defined when n is prime!).

  - $2/3 \pmod{5} \equiv 4 \pmod{5}$ (as $3 \times 4 \equiv 2 \pmod{5}$)

Results $b \pmod{n}$ are usually expressed using numbers $b \in \{0, 1, \ldots, n - 1\}$.

Cryptography: Vocabulary and Notations

- plaintext **m**: string of characters from an alphabet A ($m \in M$)

  - typically A..Z, printable ASCII, 0 and 1, 0,..., 9, etc.

- **f** encrypting algorithm - converts **m** to cyphertext $c \in C$

  - typically **f** considered public knowledge (bribery, moles, etc)

- encrypting key $e \in E$ (knowledge of this may be restricted)
- encrypt equation **c = f(e,m)**

  - To avoid ambiguity: $m \neq m' \Rightarrow c = f(e,m) \neq c' = f(e,m')$

- **g** decrypting algorithm - converts **c** to **m**

  - considered public knowledge

- used a decrypt key $\mathbf{d} \in D$ (knowledge of this may be restricted)
- for each $\mathbf{e} \in E$ there **must** be $\mathbf{d} \in D$ such that $\mathbf{m = g(d, c) = g(d, f\,(e, m))}$
- Usually d and e are related: $\mathbf{d = h(e)}$.

Caesar Cypher

- Alphabet: $A = \{A, \dots, Z\}$ so $n = 26$ (or numbers 0.25).
- m and c are single characters
- encrypt key $e \in \{0, \dots, 25\}$
- encrypt equation: $c = f(e, m) = m + e$ (mod 26)
- decrypt equation: $m = g(d, c) = c + d$ (mod 26) $(f = g)$
- clearly $d + e = 0$ (mod 26) (both need to be kept secret!)
- Method:

$$g(d, c) = g(d, f(e, m)) = g(d, m + e \quad (\text{mod } 26))$$
$$= ((m + e) \quad (\text{mod } 26)) + d) \quad (\text{mod } 26) = m$$

- Example: $e = 23$ and $n = 26$, $d = 3$
- look-up table $(e = 3)$:

$$plaintext : \quad A, B, C, D, \dots$$
$$cyphertext : \quad D, E, F, G, \dots$$

RSA Algorithm (1977)

This is an efficient algorithm which uses the properties of large primes, as well as Euler's totient function to produce encryption and decryption key.
The RSA algorithm involves: Key generation, Encryption, Decryption.

Step 1. Key generation

1. Choose two distinct prime numbers p and q.
2. Compute $n = pq$.
3. Compute $\varphi(n) = (p - 1)(q - 1)$, where φ is Euler's totient function.
4. Choose an integer e such that $1 < e < \varphi(n)$ and $\gcd(e, \varphi(n)) = 1$

- e is released as the **public key** exponent.

5. Determine d as: $d \cdot e = 1$ (mod $\varphi(n)$)

- d is kept as the **private key** exponent.

Remark: p, q and $\varphi(n)$ must be kept secret (they give d).

Step 2. Encryption

John sends his public key (n, e) to Ana and keeps the private key secret. Ana then sends message M to John. She first turns M into an integer m s.t. $0 \leq m < n$ by using a padding scheme (agreed-upon reversible protocol). She then computes the cyphertext corresponding to

$$c = m^e \pmod{n}.$$

Ana then transmits c to John.

Step 3. Decryption

John can recover m from c using his private key exponent d by computing

$$m = c^d \pmod{n}.$$

Given m, he recovers the original message M by reversing the padding scheme.

Summary

"A user of RSA creates and then publishes the product of two large prime numbers, along with an auxiliary value, as their public key. The prime factors must be kept secret. Anyone can use the public key to encrypt a message, but with currently published methods, if the public key is large enough, only someone with knowledge of the prime factors can feasibly decode the message" (Rivest: 1978).

10.2 Essential Problems

E 10.1. Find 5 pairs of Pythagorean triples $a, b, c \in \mathbb{N}$ s.t. $a^2 = b^2 + c^2$.

E 10.2. Factorize the numbers 96, 144, 286, 777 and 1001.

E 10.3. Find gcd(144, 96), gcd(1001, 777) and gcd(1001, 286).

E 10.4. Find lcm(144, 96), lcm(1001, 777) and lcm(1001, 286).

E 10.5. Compute (find results in the set $\{0, \ldots, n-1\}$ for $\pmod{n}$)):

$$5 + 2 \pmod{2}, \quad 3 \times 5 \pmod{2}$$
$$3 + 7 \pmod{5}, \quad 3 - 7 \pmod{5}, \quad 3 \times 7 \pmod{5}.$$
$$3 + 7 \pmod{11}, \quad 3 - 7 \pmod{1}1, \quad 3 \times 7 \pmod{1}1.$$

E 10.6. Find the $\varphi(n)$ function for the numbers $n = 11, 21, 24, 36, 49, 81, 100$.

E 10.7. Code the message "NOBLI EXS" using a Caesar cipher of key $e = 16$.

10.3 Supplementary Problems

S 10.1. Find the largest prime number smaller than 1000.

S 10.2. Goldbach conjecture:
For n even there are a, b primes s.t. $n = a + b$. Check this results for $n \leq 100$.

S 10.3. Find 7 pairs of twin primes (Their difference is 2. Ex: (3, 5), etc.).

S 10.4. Find 5 Mersenne primes of the form $M_p = 2^p - 1$.

S 10.5. Factorize the numbers $703, 779, 968, 1002, 1440, 1547, 1763, 2261$.

S 10.6. Find $\gcd(1440, 968)$, $\gcd(1002, 703)$, $\gcd(2261, 1547)$ and $\gcd(779, 1763)$.

S 10.7. Find $\text{lcm}(1440, 968)$, $\text{lcm}(1002, 703)$, $\text{lcm}(2261, 1547)$ and $\gcd(779, 1763)$.

S 10.8. Solve the following equations (in the set $\{0, \ldots, n - 1\}$ for $(\text{mod } n)$):

$$1 + x = 0 \quad (\text{mod } 2), \quad 3 \times x = 1 \quad (\text{mod } 2)$$
$$3 + x = 1 \quad (\text{mod } 8), \quad 3 - x = 6 \quad (\text{mod } 8), \quad 4 \times x = 7 \quad (\text{mod } 8).$$
$$5 + x = 3 \quad (\text{mod } 11), \quad 3 - x = 7 \quad (\text{mod } 11), \quad 4 \times x = 8 \quad (\text{mod } 11).$$

S 10.9. Find the $\varphi(n)$ function for the numbers $n = 19, 31, 28, 48, 144, 169, 1001$.

S 10.10. Find the RSA key d for the public key $e = 5$, for $n = 65$.

10.4 Problem Answers

Essential Problems

E 10.1. $(3, 4, 5), (5, 12, 13), (7, 24, 25), (8, 15, 17), (9, 40, 41)$.

E 10.2. The factorisations are

$96 = 2^5 \cdot 3^1$; $144 = 2^4 \cdot 3^2$; $286 = 2 \cdot 11 \cdot 13$; $777 = 3 \cdot 7 \cdot 37$; $1001 = 7 \cdot 11 \cdot 13$.

E 10.3. The greatest common divisors are

$\gcd(96, 144) = 2^4 \cdot 3 = 48$; $\gcd(1001, 777) = 7$; $\gcd(1001, 286) = 11 \cdot 13 = 143$.

E 10.4. The least common multiples are

$\text{lcm}(96, 144) = 2^5 \cdot 3^2 = 288$; $\text{lcm}(1001, 777) = 3 \cdot 7 \cdot 11 \cdot 13 \cdot 37 = 111111$;
$\text{lcm}(1001, 286) = 2 \cdot 7 \cdot 11 \cdot 13 = 2002$.

E 10.5. The modulo operations give the following results:

$$1 \quad (\text{mod } 2), \quad 1 \quad (\text{mod } 2)$$
$$0 \quad (\text{mod } 5), \quad 1 \quad (\text{mod } 5), \quad 1 \quad (\text{mod } 5).$$
$$10 \quad (\text{mod } 11), \quad 7 \quad (\text{mod } 11), \quad 10 \quad (\text{mod } 11).$$

E 10.6. One has to factorise the numbers and apply the multiplicity of $\varphi(n)$.

$$\varphi(11) = 10;$$
$$\varphi(21) = \varphi(3 \cdot 7) = \varphi(3) \cdot \varphi(7) = 2 \cdot 6 = 12;$$
$$\varphi(24) = \varphi(2^3 \cdot 3) = \varphi(2^3) \cdot \varphi(3) = 4 \cdot 2 = 8;$$
$$\varphi(36) = \varphi(2^2 \cdot 3^2) = \varphi(2^2) \cdot \varphi(3^2) = 2 \cdot 6 = 12;$$
$$\varphi(49) = \varphi(7^2) = 49(1 - 1/7) = 42;$$
$$\varphi(81) = \varphi(3^4) = 81(1 - 1/3) = 54;$$
$$\varphi(100) = \varphi(2^2 \cdot 5^2) = \varphi(2^2) \cdot \varphi(5^2) = 2 \cdot 20 = 40.$$

E 10.7. Each letter in the string is shifted to the right by 16 positions.

If the transformed letter is after Z, we use the order:

$A, B, C, D, \ldots, X, Y, Z, A, B, C, \ldots$.
Finally we obtain "DERBY UNI".

Supplementary Problems

S 10.1. By consulting a prime numbers list on finds the number 997.

S 10.2. The verification of the Goldbach conjecture yields the following pairs

$$4 = 2 + 2; \quad 6 = 3 + 3; \quad 8 = 3 + 5; \quad 10 = 3 + 7; \quad 12 = 5 + 7; \quad 14 = 3 + 11;$$
$$18 = 5 + 13; \quad 20 = 3 + 17; \quad \ldots; \quad 96 = 7 + 89; \quad 98 = 19 + 79; \quad 100 = 3 + 97.$$

S 10.3. The first few twin primes are

$$(3, 5), (5, 7), (11, 13), (17, 19), (29, 31), (41, 43), (59, 61), (71, 73), (101, 103).$$

S 10.4. The first five Mersenne primes are 3, 7, 31, 127, 8191.

S 10.5. The prime factorisations are

$$703 = 19 \cdot 37; \qquad\qquad\qquad 779 = 19 \cdot 41;$$

$$968 = 2^3 \cdot 11^2; \qquad\qquad 1002 = 2 \cdot 3 \cdot 167;$$
$$1440 = 2^5 \cdot 3^2 \cdot 5; \qquad\qquad 1547 = 7 \cdot 13 \cdot 17;$$
$$1763 = 41 \cdot 43; \qquad\qquad 2261 = 7 \cdot 17 \cdot 19.$$

S 10.6. The greatest common divisors are

$$\gcd(1440, 968) = 2^3 = 8;$$
$$\gcd(1002, 703) = 1;$$
$$\gcd(2261, 1547) = 7 \cdot 17 = 119;$$
$$\gcd(779, 1763) = 41.$$

S 10.7. The least common multiples are

$$\mathrm{lcm}(1440, 968) = 2^5 \cdot 3^2 \cdot 5 \cdot 11^2;$$
$$\mathrm{lcm}(1002, 703) = 2 \cdot 3 \cdot 19 \cdot 37 \cdot 167;$$
$$\mathrm{lcm}(2261, 1547) = 7 \cdot 13 \cdot 17 \cdot 19;$$
$$\mathrm{lcm}(779, 1763) = 19 \cdot 41 \cdot 43.$$

S 10.8. Using definitions of the modulo operations one obtains:

$$x = 1 \quad (\mathrm{mod}\ 2), \quad x = 1 \quad (\mathrm{mod}\ 2)$$
$$x = 6 \quad (\mathrm{mod}\ 8), \quad x = 5 \quad (\mathrm{mod}\ 8), \quad \text{there is no such } x.$$
$$x = 9 \quad (\mathrm{mod}\ 11), \quad x = 7 \quad (\mathrm{mod}\ 11), \quad x = 2 \quad (\mathrm{mod}\ 11).$$

S 10.9. Using the formula for $\varphi(n)$ one obtains

$$\varphi(19) = 18; \qquad \varphi(31) = 30; \qquad \varphi(28) = 12; \qquad \varphi(48) = 16;$$
$$\varphi(144) = 36; \qquad \varphi(169) = 156; \qquad \varphi(1001) = 720.$$

S 10.10. $n = 65$ then $p = 5, q = 13$ and $\varphi(65) = 48$. To find d one needs to solve $5 * d = 1$ (mod 48) which gives $d = 30$.

Chapter 11
Elements of Calculus

Abstract Calculus is one of the important subjects of mathematics, with applications in various domains of knowledge such as Finance, Engineering or Architecture. The major branches of Calculus are differential calculus (concerning rates of change and slopes of curves), and integral calculus (concerning accumulation of quantities and the areas under curves), linked together through the Fundamental Theorem of Calculus. In this Chapter we present some key elements of Calculus such as sequences, limits, convergence, as well as definitions and rules for the differentiation and integration of basic functions such as polynomials and basic trigonometric functions.

Keywords Sequences · Limits · Continuity · Derivative · Integral

11.1 Brief Theoretical Background

Sequences and Limits

Sequence: ordered list of objects. Discrete function $x : \mathbb{N} \to X$, $x(n) = x_n$.

Limit of a sequence: sequence terms **"eventually get close to"** a value.

- We say that the sequence **converges** to the limit
- If a sequence does not converge it is **divergent**

Examples: The following sequences have the limits:

- even numbers: $2, 4, 6, ..., 2n$ (diverge)
- inverse numbers: $1, \frac{1}{2}, \frac{1}{3}, ..., \frac{1}{n}, ...$ (converge to zero)
- geometric progression: $1, \frac{1}{2}, \frac{1}{2^2}, ..., \frac{1}{2^n}, ...$ (converges to zero)
- $0.3, 0.33, 0.333, 0.3333, ...$ converges to $1/3$.

Limit of a function: Consider the function $f : R \to R$ and numbers $p, L \in R$.

O. Bagdasar, *Concise Computer Mathematics*, SpringerBriefs in Computer Science, DOI: 10.1007/978-3-319-01751-8_11, © The Author(s) 2013

We say "the limit of f as x approaches p is L" and write

$$\lim_{x \to p} f(x) = L,$$

if for every convergent sequence $x_n \to p$ we have $f(x_n) \to L$.
Examples: The following functions have the limits:

- $f(x) = x^2 \quad \lim_{x \to 2} f(x) = 2^2 = 4$.
- $f(x) = 1/x \quad \lim_{x \to 0} f(x) = \infty$.
- $f(x) = \sin x \quad \lim_{x \to 0} \sin x = 0$.

Continuous Functions

A function $f : \mathbb{R} \to \mathbb{R}$ is continuous if

- for every point x and convergent sequence $x_n \to x$ we have $f(x_n) \to f(x)$.
- the graph is a single unbroken curve with no "holes" or "jumps".

The intermediate value theorem: If f is continuous on $[a, b]$ and $f(a)f(b) < 0$ then there is $c \in [a, b]$ s.t. $f(c) = 0$.

Examples: Polynomials, exp, sin, cos are all continuous functions.
The cubic in Fig. 11.1a crosses line $y = 0$ between -2 and 0, changing sign.
Signum (has the values ± 1) (Fig. 11.1b) changes sign but is not continuous.

The Derivative

Measure of how a function changes as its input changes.
For x and $x + h$ we define the ratio

$$m = \frac{\Delta f(x)}{\Delta x} = \frac{f(x + h) - f(x)}{(x + h) - (x)} = \frac{f(x + h) - f(x)}{h}.$$

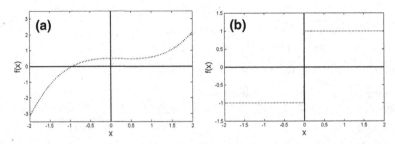

Fig. 11.1 a $f(x) = x^3/3 - x^2/4 + 1/2$; **b** $f(x) = \text{sgn}(x)$

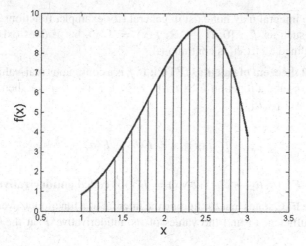

Fig. 11.2 Graph of function $f(x) = x^3 \sin(x)$ between $a = 1$ and $b = 3$

The derivative of the function $f : \mathbb{R} \to \mathbb{R}$ at x is the limit

$$f'(x) = \lim_{h \to 0} \frac{f(x+h) - f(x)}{h}$$

Differentiation of elementary functions

- powers: $f(x) = x^n (n \in \mathbb{N})$: $f'(x) = nx^{n-1}$
- polynomials: $(ax^2 + bx + c)' = 2ax + b$ (a, b, c: constants)
- trigonometric functions:

$$f(x) = \sin(x) : \quad f'(x) = \cos(x)$$
$$f(x) = \cos(x) : \quad f'(x) = -\sin(x)$$

The Definite Integral

Let the function f defined on $[a, b]$. The definite integral denoted by

$$\int_a^b f(x)\, dx$$

represents the area of the region in the xy-plane bounded by:

- the graph of f
- the x-axis,
- vertical lines $x = a$ and $x = b$.

Remark: The integral may not exist in general (for example, functions which have singularities, such as $f : [0, 1] \rightarrow \mathbb{R}, f(x) = 1/x$), but it does exist when the function f defined on $[a, b]$ is continuous!

Fundamental theorem of calculus (FTC): If f is a continuous real-valued function defined on $[a, b]$ and a function $F : [a, b] \rightarrow \mathbb{R}$ s.t. $F'(x) = f(x)$, then the definite integral of f over $[a, b]$ is

$$\int_a^b f(x)\,dx = F(b) - F(a)$$

The function $F : [a, b] \rightarrow \mathbb{R}$ s.t. $F'(x) = f(x)$ is called **antiderivative**.

Remark: The FTC states that for finding the integral of a function f over an interval $[a, b]$, it is sufficient to find the values of its antiderivative F at the ends of the interval!

Integration of elementary functions

- powers: $f(x) = x^n (n \in \mathbb{N})$: $F(x) = \int f(x)dx = \frac{x^{n+1}}{n+1} + C$ (constant)
- polynomials: $f(x) = ax + b$; $F(x) = \int f(x)dx = a\frac{x^2}{2} + bx + C$
  (a, b, C: constants)
- trigonometric functions:

$$f(x) = \sin(x) : \quad F(x) = \int f(x)dx = -\cos(x) + C$$

$$f(x) = \cos(x) : \quad F(x) = \int f(x)dx = \sin(x) + C$$

11.2 Essential Problems

E 11.1. Find the limit of the following sequences

(1) $x_n = 1$;   (2) $x_n = 1 + 1/n$;   (3) $x_n = 1/n^2$;   (4) $x_n = 1/2^n$;
(5) $0.9, 0.99, 0.999, 0.9999, \ldots$;
(6) $0.23, 0.23, 0.2323, 0.232323, \ldots$;
(7) $0.423, 0.42323, 0.4232323, 0.423232323, \ldots$

E 11.2. Prove that $f(x) = 2x^2 + 2x - 5$ has a root in $[1, 2]$ and $[-3, -2]$.

E 11.3. Prove that the cubic $f(x) = 2x^3 + 2x^2 - 3$ has a root $x_0 \in [0, 1]$.

E 11.4. Differentiate the functions: (1) $f(x) = 2x + 1$; (2) $f(x) = 2x^2 + 3$.

E 11.5. Integrate: (1) $f(x) = 1$; (2) $f(x) = 2x + 3$; (3) $f(x) = 2x^2 + 3$.

E 11.6. Evaluate $\int_0^2 f(x)dx$ for $f(x) = x^2 + 2$ using the **FTC**.

E 11.7. Find the area between the curves $f(x) = -x^2 + 6x - 2, g(x) = 2x + 1$.

11.3 Supplementary Problems

S 11.1. Find the limit of the following sequences

$$(1) \quad x_n = \frac{2n}{n-4};$$

$$(2) \quad x_n = (-1)^n \frac{2^n}{n!};$$

$$(3) \quad x_n = n^2 - 5n;$$

$$(4) \quad x_n = \sqrt{n+1} - \sqrt{n}.$$

S 11.2. How many roots does the cubic $f(x) = 8x^3 + 12x^2 - 2x - 3$ have?

S 11.3. Prove that $f(x) = x^2 - 3\cos\left(\frac{\pi}{2}x\right) + 2$ has a root $x_0 \in [0, 1]$.

S 11.4. Differentiate the function $f(x) = 2x^3 + 7x + 3$.

S 11.5. Integrate: (1) $f(x) = 2x + 7\cos(x) + 3\sin(x)$; (2) $f(x) = 2x^3 + 1$.

S 11.6. Consider $f(x) = -10x^2 + 50x - 40$ and $g(x) = 5x - x^2$. Find

(a) the area under $f(x)$ and above $y = 0$;
(b) the area under $g(x)$ and above $y = 0$;
(c) the $x_1 < x_2$ values where the curves $f(x)$ and $g(x)$ intersect;
(d) the area between $f(x)$ and $g(x)$.

11.4 Problem Answers

Essential problems

E 11.1. (1) 0; (2) 0; (3) 0; (4) 0; (5) 1; (6) 23/99; (7) 423/990.

E 11.2. $f(1)f(2) = (-1)7 < 0$; $f(-3)f(-2) = (7)(-1) < 0$.

E 11.3. $f(0)f(1) = (-3)1 < 0$.

E 11.4. (1) $f'(x) = 2$; (2) $f'(x) = 4x$.

E 11.5. (1) $\int f(x)dx = x + C$; (2) $\int f(x)dx = x^2 + 3x + C$; (3) $\int f(x)dx = \frac{2x^3}{3} + 3x + C$.

E 11.6. $F(x) = \int f(x)dx = \frac{x^3}{3} + 2x + C$. The integral is therefore given by

$$\int_0^2 f(x)dx = F(2) - F(0) = 8/3 + 4 + C - C = 20/3.$$

E 11.7. We first solve $f(x) = g(x)$ and obtain the solutions $x_1 = 1$, $x_2 = 3$.
$A = \int_1^3 f(x) - g(x)dx = \int_1^3 -(x^2 - 4x + 3)dx = -(x^3/3 - 4x^2 + 3x + C)|_1^3 = 4/3.$

Supplementary problems

S 11.1. (1) 2;   (2) 0;   (3) ∞;   (4) 0.

S 11.2. $f(-2) = -11$; $f(-1) = 3$; $f(0) = -3$; $f(1) = 15$. We have $f(-2)f(-1) < 0$, $f(-1)f(0) < 0$ and $f(0)f(1) < 0$.

S 11.3. $f(0)f(1) = (-1)3 < 0$.

S 11.4. $f'(x) = 6x^2 + 7$.

S 11.5. (1) $\int f(x)dx = x^2 + 7\sin(x) - 3\cos(x)$; (2) $\int f(x)dx = x^4/2 + C$.

S 11.6. (a) 45.00;   (b) 125/6 = 20.83;   (c) $x_1 = 1/6(15 - \sqrt{65}) = 1.15$, $x_2 = 1/6(15 + \sqrt{65})$;   (d) $\int_{1/6(15-\sqrt{65})}^{1/6(15+\sqrt{65})}(-10(x - 1)(x - 4) + x(x - 5))dx = \frac{65\sqrt{65}}{18} \sim 29.1137$.

Chapter 12
Elementary Numerical Methods

Abstract Very often in real applications, one has to solve some complicated equations, where a formula for the result is either impossible to obtain, or it is too complicated for any practical use. In this case one has to rely on a wide range of numerical methods, which very often represent approximations to the real results. These methods are very important in practice and usually offer both an algorithm generating increasingly exact approximations and an approximation for the error. In this chapter we shall discuss the Lagrange polynomial for interpolation data given by a table of values, basic iterative methods used for the numerical integration of real functions and some iterative methods for root finding.

Keywords Interpolation · Trapezium rule · Bisection method · Newton-Raphson

12.1 Brief Theoretical Background

In this section we present the basic basic notions regarding polynomial interpolation, numerical integration and root finding algorithms.

Polynomial Interpolation

When working with data on given by (x, y) coordinates, it is often required to find a curve passing through multiple points. Polynomials are usually the first choice for such a curve, as they are smooth, easy to work with and have a formula. The basic idea of the Lagrange polynomial is to fit a minimum degree polynomial to a set of points.

O. Bagdasar, *Concise Computer Mathematics*, SpringerBriefs in Computer Science,
DOI: 10.1007/978-3-319-01751-8_12, © The Author(s) 2013

Lagrange Polynomial

For $n \geq 2$ and a set of points $P_1(x_1, y_1), P_2(x_2, y_2), \ldots, P_n(x_n, y_n)$ s.t. $x_1 < x_2 < \cdots < x_n$ one can find a polynomial function $L(x)$ of degree $n - 1$ passing through the points $P_1, \ldots, P_n$ (i.e. $L(x_1) = y_1, L(x_2) = y_2, \ldots, L(x_n) = y_n$).

Examples: Given the points $P_1(1, 2), P_2(2, 5)$ and $P_3(3, 3)$ we can:
1. Find a polynomial passing through point P_1.
Solution: This could be any polynomial $L(x)$ for which $L(1) = 2$. The polynomial of minimum degree is the constant polynomial $L(x) = 2$.
2. Find a polynomial passing through points P_1 and P_2.
Solution: The interpolating polynomial is:

$$L(x) = 2 \cdot \frac{x - 2}{1 - 2} + 5 \cdot \frac{x - 1}{2 - 1}$$
$$= -2(x - 2) + 5(x - 1) = 3x - 1.$$

3. Find a polynomial passing through point P_1, P_2 and P_3.
Solution: The interpolating polynomial is:

$$L(x) = 1 \cdot \frac{x - 2}{1 - 2} \cdot \frac{x - 3}{1 - 3} + 5 \cdot \frac{x - 1}{2 - 1} \cdot \frac{x - 3}{2 - 3} + 3 \cdot \frac{x - 1}{3 - 1} \cdot \frac{x - 2}{3 - 2}$$
$$= -3x^2 + 13x - 9.$$

In general, one finds the polynomials $L_1, \ldots, L_n$ such that

$$L(x) = L_1(x)y_1 + L_2(x)y_2 + \cdots + L_n(x)y_n,$$
$$= a_{n-1}x^{n-1} + a_{n-2}x^{n-2} + a_1 x + a_0$$

which satisfy

$$L_j(x) = \begin{cases} 1, & x = x_j \\ 0, & x \in \{x_1, \ldots, x_n\} \setminus \{x_j\}. \end{cases}$$

The degree of the polynomial is justified by the fact that a polynomial of degree $n - 1$ depends on n coefficients, equal to the number of available points.

Numerical Integration

The problem of finding the area between the graph of the function $f : [a, b] \to \mathbb{R}$ and the horizontal axes was linked in Chap. 11 to evaluating the integral $\int_a^b f(x)dx$. However, in real applications the functions to be integrated are very complicated or the function might be known at some discrete points of the interval $[a, b]$. In such cases, the area is approximated by numerical integration methods.

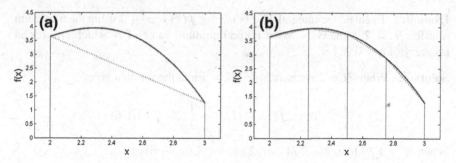

Fig. 12.1 Approximations for $f(x) = x^2 \sin x$ using (**a**) $N = 1$; (**b**) $N = 4$ intervals

Trapezium rule
The graph of function $f : [a, b] \to \mathbb{R}$ is approximated by a linear function (Fig. 12.1)

$$\int_a^b f(x)\, dx \approx (b - a)\frac{f(a) + f(b)}{2}$$

Composite trapezium rule (iterative method): To improve accuracy, the interval $[a, b]$ can be divided into N smaller intervals where the trapezium rule is applied. The following notions are defined:

- **Step:** $h = \frac{b-a}{N}$ (step);
- **Partition:** $x_0 = a$, $x_1 = a + h$, $x_2 = a + 2h$,..., $x_N = a + Nh = b$.

We apply the trapezium rule and obtain the following approximation for the area

$$I = \int_a^b f(x)\, dx \approx T_N = \frac{h}{2}\sum_{k=1}^{N}(f(x_{k-1}) + f(x_k))$$

$$= \frac{b-a}{2N}(f(x_0) + 2f(x_1) + 2f(x_2) + \ldots + 2f(x_{N-1}) + f(x_N)).$$

Error bound: The absolute error E_{T_N} of this approximation is bounded by

$$E_{T_N} = |I - T_N| \leq \frac{(b-a)h^2}{12}M = \frac{(b-a)^3}{12N^2}M \quad (M = \max_{\xi \in [a,b]} |f''(\xi)|).$$

Number of intervals needed: If the desired accuracy is ε (say 0.01) the number of intervals needed satisfies the inequality

$$N \geq \sqrt{\frac{(b-a)^3 M}{12\varepsilon}}.$$

Example: Evaluate the integral $\int_0^2 f(x)dx$ for $f(x) = x^2 + 2$ using the trapezium rule for $N = 2, 4$ intervals. What is the minimum value of N which ensures an accuracy of $\varepsilon = 0.01$?

Solution: When $N = 2$ we have $h = 2/2 = 1$ and the formula gives

$$T_2 = \frac{1}{2}(f(0) + 2f(1) + f(2)) = \frac{1}{2}(2 + 2\cdot 3 + 6) = 7.$$

When $N = 4$ we have $h = 2/4 = 1/2$ and the formula gives

$$T_4 = \frac{1}{4}(f(0) + 2f\left(\frac{1}{2}\right) + 2f(1) + 2f\left(\frac{3}{2}\right) + f(2))$$

$$= \frac{1}{4}(2 + 2\frac{9}{4} + 2\cdot 3 + 2\frac{17}{4} + 6) = 27/4.$$

The antiderivative of f is $F(x) = \int f(x)dx = \frac{x^3}{3} + 2x + C$ therefore

$$I = \int_0^2 f(x)dx = F(2) - F(0) = 8/3 + 4 + C - C = 20/3.$$

The error analysis for this problem yields

- The absolute error is $E_{T_N} = |I - T_N|$. For $N = 2$, $E_{T_2} = |20/3 - 7| = 1/3$.
- The error bound is $E_B = \frac{(b-a)h^2}{12} \max_{\xi\in[a,b]} |f''(\xi)|$.
  In this example $f''(x) = 2$, therefore $M = \max_{\xi\in[0,2]} |f''(\xi)| = 2$.

For $N = 2$, $h = (b-a)/N = 1$, so $E_{B_2} = \frac{2\cdot 1^2}{12}\cdot 2 = 1/3$.
For $N = 4$, $h = (b-a)/N = 0.5$, so $E_{B_2} = \frac{2\cdot 0.5^2}{12}\cdot 2 = 1/12$.

As $f''(x) = 2$ (so $M = 2$), the number N of intervals required is given by the formula

$$N \geq \sqrt{\frac{(2-0)^3 \cdot 2}{12 \cdot 0.01}} \sim 11.53, \text{ therefore } N = 12.$$

Root Finding Algorithms

Few equations can be solved by a simple formula and a numerical method is often required to approximate the solution. Many times, these approximations can be improved by successive iterations. In this section we shall briefly present the bisection method, along with the Newton-Rapson and Secant iterations.

Root finding: The problem of finding $f(x) = 0$, where $f : \mathbb{R} \to \mathbb{R}$.

Examples: Find the roots of the equations $f(x) = 0$:

$$f(x) = x^2 - 4x + 3$$
$$f(x) = x^3 - 4x + 3$$
$$f(x) = \sin(3x) - \sin(x)$$
$$f(x) = e^x - 5\sin(3x) + x^2$$

The usual starting point is to plot the graph of the function and then guess...

Corollary of the intermediate value theorem If f is continuous on $[a, b]$ and $f(a)f(b) < 0$ then there is $c \in [a, b]$ s.t. $f(c) = 0$.

Bisection method algorithm

- **Inputs:** $f(x)$ (function), interval $[a_0, b_0]$ s.t. $f(a_0)f(b_0) < 0$
- **Output:** An approximation of the root of $f(x) = 0$ in $[a_0, b_0]$.
- **Algorithm** For $k = 0, 1, 2, ...$, do until satisfied:

  - Compute $c_k = \frac{a_k + b_k}{2}$.
  - Test if c_k is the desired root, if so, stop.
  - If c_k is not the desired root, test if $f(c_k)f(a_k) < 0$

    (1) If so, set $b_{k+1} = c_k$ and $a_{k+1} = a_k$.
    (2) Otherwise, set $a_{k+1} = c_k$ and $b_{k+1} = b$.

Bisection method example
Let $f(x) = x^3 - 6x^2 + 11x - 6$. Let $a_0 = 2.5, b_0 = 4$. Then $f(a_0)f(b_0) < 0$ so there is a root in $[2.5, 4]$.
Iteration 1. $k = 0$:
$$c_0 = \frac{a_0 + b_0}{2} = \frac{2.5 + 4}{2} = 3.25.$$

Since $f(a_0)f(c_0) = f(2.5)f(3.25) < 0$, set $a_1 = a_0$ and $b_1 = c_0$.
Iteration 2. $k = 1$:
$$c_1 = \frac{a_1 + b_1}{2} = \frac{2.5 + 3.25}{2} = 2.875.$$

Since $f(a_1)f(c_1) = f(2.5)f(3.25) > 0$, set $a_2 = c_1$ and $b_2 = b_1$.
Iteration 3. $k = 2$:
$$c_2 = \frac{a_2 + b_2}{2} = \frac{2.875 + 3.250}{2} = 3.0625.$$

Since $f(a_2)f(c_2) = f(2.875)f(3.250) < 0$, set $a_3 = a_2$ and $b_2 = c_2$. It is clear that the iterations are converging towards the root $x = 3$.

Table 12.1 Successive approximations using Newton's algorithm

Iteration	x_n	$f(x_n)$	$f'(x_n)$
0	0.25	0.125	5
1	0.225	0.00125	4.9
2	0.224745	0.0000001	4.89898
3	0.224745	0	4.898979

Bisection method: number of iterations needed

Starting from the initial interval $[a_0, b_0]$, the minimum number of iterations which ensure that

$$|c_n - x^*| \leq \frac{b_n - a_n}{2} = \frac{b_0 - a_0}{2^{n+1}} \leq \varepsilon$$

is give by the formula

$$n \geq \frac{\log(b - a) - \log(2\varepsilon)}{\log 2}.$$

Examples: Suppose we want to know a priori the minimum number of iterations needed in the bisection algorithm, for $a_0 = 2.5$, $b_0 = 4$ and $\varepsilon = 10^{-3}$.

$$n \geq \frac{\log(1.5) - \log(0.002)}{\log 2} = 9.5507 \Longrightarrow n = 10.$$

Newton's Method

This is an often used method, with good accuracy and having the elements

- **Starting point:** x_0 (the choice of this point is important)
- **Iteration:**

$$x_{n+1} = x_n - \frac{f(x_n)}{f'(x_n)}$$

Examples: Find the roots of the equation $f(x) = 2x^2 + 4x - 1 = 0$. We first compute $f'(x) = 4x + 4$ and then write the iteration formula

$$x_{n+1} = x_n - \frac{2x_n^2 + 4x_n - 1}{4x_n + 4},$$

Starting from $x_0 = 0.25$ we obtain

Remark: Different roots may be found by starting from different values.

12.2 Essential Problems

Polynomial Interpolation

E 12.1. Find the Lagrange polynomial interpolating the data

$$
\begin{array}{ll}
x_1 = 1 & x_2 = 2 \\
y_1 = 1 & y_2 = 3
\end{array}
$$

E 12.2. Find the Lagrange polynomial interpolating the data

$$
\begin{array}{lll}
x_1 = 1 & x_2 = 2 & x_3 = 3 \\
y_1 = 1 & y_2 = 2 & y_3 = 5
\end{array}
$$

Numerical Integration

E 12.3. Evaluate $\int_0^2 f(x)dx$ for $f(x) = x^2 + 2$ using $N = 2$ or 4 trapeziums.

E 12.4. Evaluate the integral $\int_0^2 f(x)dx$ for $f(x) = x^3 + 4x^2 - 5x - 2$ exactly and by the trapezium rule for 2, 4 and 8 intervals. What can you notice?

E 12.5. Evaluate the integral $\int_0^1 f(x)dx$ for $f(x) = 4x^2 - 5x - 2$ exactly and by the trapezium rule for $N = 2, 4, 8$. What is the error in each case?

E 12.6. Evaluate the composite trapezium approximation for the integral

$$
\int_0^{\pi/2} x^2 \cos x\, dx
$$

(a) with $N = 2$ intervals;   (b) with $N = 4$ intervals.

Root Finding

E 12.7. (Existence of roots)

(a) Find the number of roots for the cubic $f(x) = 8x^3 + 12x^2 - 2x - 3$.
(b) Prove that $f(x) = x^2 - 3\cos\left(\frac{\pi}{2}x\right) + 2$ has a root $x_0 \in [0, 1]$.

E 12.8. (The bisection algorithm) (a) Apply 4 iterations of the bisection algorithm (c_0, c_1, c_2, c_3) to the function

$$
f(x) = 2x^3 + 2x^2 - 3, \quad x \in [0, 1].
$$

(b) How many iterations are needed to ensure a precision of 10^{-4} ?

E 12.9. (Newton's algorithm) Apply 3 iterations of Newton's method (find x_1, x_2, x_3) for the function

$$
f(x) = 6x^2 + x - 2,
$$

starting from the points (a) $x_0 = -1$; (b) $x_0 = 2$ and find the exact roots.

12.3 Supplementary Problems

S 12.1. Find the Lagrange interpolation polynomial for the coordinates

$$x_1 = 2 \qquad\qquad\qquad\qquad x_2 = 3$$
$$y_1 = 1 \qquad\qquad\qquad\qquad y_2 = 5$$

S 12.2. Find the Lagrange interpolation polynomial for the coordinates

$$x_1 = 1 \qquad\qquad x_2 = 3 \qquad\qquad x_3 = 4,$$
$$y_1 = -5 \qquad\qquad y_2 = -1 \qquad\qquad y_3 = 10,$$

S 12.3. Evaluate the integral $\int_{-1}^{1} f(x)dx$ for $f(x) = x^2 - x + 1$ exactly and by using the trapezium rule for $N = 2, 4, 8$. Which value of N ensures an accuracy of 0.01?

S 12.4. Evaluate the integral $\int_{0}^{\pi} f(x)dx$ for $f(x) = \sin x$ exactly and by using the trapezium rule for $N = 2, 4, 8$. How many intervals N are required for ensuring an accuracy of 0.01? But for an accuracy of 0.0001?

S 12.5. (Limitations of the bisection algorithm) Apply a few iterations of the bisection algorithm on the interval $[0, 2]$ to

$$\text{(i) } f(x) = \frac{1}{x-1}; \qquad\qquad \text{(ii) } g(x) = x^3 - x^2 - x + 1.$$

How many iterations are needed to ensure a precision of 10^{-4}?

12.4 Problem Answers

Essential Problems

E 12.1. The interpolating polynomial is:

$$L(x) = 1 \cdot \frac{x-2}{1-2} + 3 \cdot \frac{x-1}{2-1} = -(x-2) + 3(x-1) = 2x - 1.$$

E 12.2. The interpolating polynomial is:

$$L(x) = 1 \cdot \frac{x-2}{1-2} \cdot \frac{x-3}{1-3} + 2 \cdot \frac{x-1}{2-1} \cdot \frac{x-3}{2-3} + 5 \cdot \frac{x-1}{3-1} \cdot \frac{x-2}{3-2}$$
$$= x^2 - 2x + 2.$$

Table 12.2 Successive approximations using Newton's algorithm

Iteration	x_n	$f(x_n)$	$f'(x_n)$	x_n	$f(x_n)$	$f'(x_n)$
0	2	24	25	−1	3	−11
1	1.04	5.5296	13.48	−0.7272	0.4462	−7.7272
2	0.6297	1.0096	8.5575	−0.6695	0.0200	−7.0342
3	0.5118	0.0835	7.1417	−0.6666	0.0000	−7.0000

E 12.3. When $N = 2$ we have $h = 2/2 = 1$, and the approximation

$$T_2 = \frac{1}{2}(f(0) + 2f(1) + f(2)) = \frac{1}{2}(2 + 2 \cdot 3 + 6) = 7.$$

When $N = 4$ we have $h = 2/4 = 1/2$, and the approximation

$$T_4 = \frac{1}{4}\left(f(0) + 2f\left(\frac{1}{2}\right) + 2f(1) + 2f\left(\frac{3}{2}\right) + f(2)\right)$$

$$= \frac{1}{4}\left(2 + 2\frac{9}{4} + 2 \cdot 3 + 2\frac{17}{4} + 6\right) = 27/4.$$

E 12.4. The composite trapezium formula generates the approximations

$$T_2 = 3; \ T_4 = 1.25; \ T_8 = 0.81; \ T_{16} = 0.70; \ T_{32} = 0.67,$$

which approach the exact value of the integral $\int_0^2 f(x)dx = 2/3$.

E 12.5. The approximations are $T_2 = 1$, $T_4 = 0.875$, $T_8 = 0.843 \sim \int_0^1 f(x)\,dx = 5/6$. The error upper bounds are $E_{T_2} = 0.1666$, $E_{T_4} = 0.0416$, $E_{T_8} = 0.0104$.

E 12.6. (a) 0.342 ; (b) 0.435.

E 12.7. (a) $f(-2) = -11$, $f(-1) = 3$, $f(0) = -3$, $f(1) = 15$ therefore we have at least three roots. As a cubic is continuous and has at most three roots, the answer is 3.

(b) The function is continuous, $f(0) = -1$ and $(1) = 3$, so $f(0)f(1) < 0$.

E 12.8. (a) $c_0 = 0.5$, $c_1 = 0.75$, $c_2 = 0.875$, $c_3 = 0.9375$; (b) $n = 13$.

E 12.9. The derivative of $f(x)$ is $f'(x) = 12x + 1$, therefore the iteration is

$$x_{n+1} = x_n - \frac{6x_n^2 + x_n - 2}{12x_n + 1}$$

The analysis suggests that the roots are $1/2$ and $-2/3$ (Table 12.1).

Supplementary Problems

S 12.1. The interpolating polynomial is:

$$L(x) = 1 \cdot \frac{x-3}{2-3} + 5 \cdot \frac{x-2}{3-2}$$
$$= -(x-3) + 5(x-2) = 4x - 7.$$

S 12.2. The interpolating polynomial is:

$$L(x) = -5 \cdot \frac{x-3}{1-3} \cdot \frac{x-4}{1-4} + -1 \cdot \frac{x-1}{3-1} \cdot \frac{x-4}{3-4} + 10 \cdot \frac{x-1}{4-1} \cdot \frac{x-3}{4-3}$$
$$= 3x^2 - 10x + 2.$$

S 12.3. The exact integral is $\int_{-1}^{1} x^2 - x + 1 dx = \frac{8}{3} \sim 2.666$. The approximations are $T_2 = 3$; $T_4 = 2.75$; $T_8 = 2.687$. The number of required intervals for a 0.01 accuracy is $N = 12$.

S 12.4. The exact integral is $\int_0^2 \sin x dx = -\cos x|_0^\pi = 2$. The approximations are $T_2 = 1.570$; $T_4 = 1.896$; $T_8 = 1.974$. $N = 17$ intervals for a 0.01 accuracy and $N = 161$ intervals for 0.0001.

S 12.5. (i) The function changes sign on $[0, 2]$ but the approximations converge to an asymptote. The method is not reliable because the function is not continuous on the interval $[0, 2]$.

(ii) The change of sign test is not relevant $f(0)f(2) = 3$, therefore the method cannot be applied. The method is not able to pick the root $x = 1$ ($g(1) = 0$).

Further Reading

1. Lipschutz, S.: Schaum's Outline of Theory and Problems of Set Theory and Related Topics, 2nd edn. McGraw-Hill, New York (1998)
2. Lipschutz, S., Lipson, M.: Schaum's Outline of Discrete Mathematics, 3rd edn. McGraw-Hill, New Year (2007)
3. Haggerty, R.: Discrete Mathematics for Computing. Addison Wesley, New York (2002)
4. Stroud, K.A., Dexter, J.B.: Foundation Mathematics. Macmillan, London (2009)
5. Tunnicliffe, W.R.: Mathematics for Programmers. Prentice Hall, Englewood Cliffs (1991)
6. Grassmann, W.K., Tremblay, J.P.: Logic and Discrete Mathematics: A Computer Science Perspective. Prentice Hall, Englewood Cliffs (1996)
7. Kelly, J.: The Essence of Logic. Prentice Hall, Englewood Cliffs (1997)
8. Neville, D.: The Essence of Discrete Mathematics. Prentice Hall, Englewood Cliffs (1997)
9. Lang, S.: Algebra. Graduate Texts in Mathematics 211, revised 3rd edn. Springer, New York (2002)
10. Lipschutz, S.: Schaum's Outline of Theory and Problems of Essential Computer Mathematics. McGraw-Hill, New York (1982)
11. Marsden, J.E.: Basic Complex Analysis, 2nd edn. W.H. Freeman, New York (1987)
12. Van Verth, J.M., Bishop, L.M.: Essential Mathematics for Games and Interactive Applications: A Programmer's Guide, 2nd edn. Elsevier Science, Amsterdam (2008)
13. Wunsch, D.A.: Complex Variables with Applications, 3rd edn. Pearson, Boston (2005)
14. Burton, D.M.: Elementary Number Theory, 7th edn. McGraw-Hill, New York (2011)
15. Churchhouse, R.: Codes and Ciphers. Cambridge University Press, Cambridge (2002)
16. Dangerfeld, J.: Decision 1 for AQA. Pearson Longman, Boston (2005)
17. Goodaire, E.G., Parmenter, M.M.: Discrete Mathematics with Graph Theory. Prentice Hall, Englewood Cliffs (1998)
18. Hill, R.: A First Course in Coding Theory. Oxford University Press, Oxford (1997)
19. Rivest, R., Shamir, A., Adleman, L.: A method for obtaining digital signatures and public-key cryptosystems. Commun. ACM **21**(2), 120–126 (1978)
20. Spivak, M.: Calculus, 4th edn. Publish or Perish, Houston (2008)
21. Stewart, J.: Calculus. Brooks/Cole, Pacific Grove (2003)

O. Bagdasar, *Concise Computer Mathematics*, SpringerBriefs in Computer Science,
DOI: 10.1007/978-3-319-01751-8, © The Author(s) 2013

Index

O. Bagdasar, *Concise Computer Mathematics*, SpringerBriefs in Computer Science,
DOI: 10.1007/978-3-319-01751-8, © The Author(s) 2013